Analog Electronics for Radiation Detection

Devices, Circuits, and Systems

Series Editor
Krzysztof Iniewski
Emerging Technologies CMOS Inc.
Vancouver, British Columbia, Canada

PUBLISHED TITLES:

Atomic Nanoscale Technology in the Nuclear Industry
Taeho Woo

Biological and Medical Sensor Technologies
Krzysztof Iniewski

Building Sensor Networks: From Design to Applications
Ioanis Nikolaidis and Krzysztof Iniewski

**Cell and Material Interface: Advances in Tissue Engineering,
Biosensor, Implant, and Imaging Technologies**
Nihal Engin Vrana

Circuits at the Nanoscale: Communications, Imaging, and Sensing
Krzysztof Iniewski

CMOS: Front-End Electronics for Radiation Sensors
Angelo Rivetti

**CMOS Time-Mode Circuits and Systems: Fundamentals
and Applications**
Fei Yuan

Design of 3D Integrated Circuits and Systems
Rohit Sharma

Electrical Solitons: Theory, Design, and Applications
David Ricketts and Donhee Ham

Electronics for Radiation Detection
Krzysztof Iniewski

Electrostatic Discharge Protection: Advances and Applications
Juin J. Liou

**Embedded and Networking Systems:
Design, Software, and Implementation**
Gul N. Khan and Krzysztof Iniewski

Energy Harvesting with Functional Materials and Microsystems
Madhu Bhaskaran, Sharath Sriram, and Krzysztof Iniewski

Gallium Nitride (GaN): Physics, Devices, and Technology
Farid Medjdoub

PUBLISHED TITLES:

Graphene, Carbon Nanotubes, and Nanostuctures:
Techniques and Applications
James E. Morris and Krzysztof Iniewski

High-Speed Devices and Circuits with THz Applications
Jung Han Choi

High-Speed Photonics Interconnects
Lukas Chrostowski and Krzysztof Iniewski

High Frequency Communication and Sensing:
Traveling-Wave Techniques
Ahmet Tekin and Ahmed Emira

Integrated Microsystems: Electronics, Photonics, and Biotechnology
Krzysztof Iniewski

Integrated Power Devices and TCAD Simulation
Yue Fu, Zhanming Li, Wai Tung Ng, and Johnny K.O. Sin

Internet Networks: Wired, Wireless, and Optical Technologies
Krzysztof Iniewski

Ionizing Radiation Effects in Electronics: From Memories to Imagers
Marta Bagatin and Simone Gerardin

Labs on Chip: Principles, Design, and Technology
Eugenio Iannone

Laser-Based Optical Detection of Explosives
Paul M. Pellegrino, Ellen L. Holthoff, and Mikella E. Farrell

Low Power Emerging Wireless Technologies
Reza Mahmoudi and Krzysztof Iniewski

Medical Imaging: Technology and Applications
Troy Farncombe and Krzysztof Iniewski

Metallic Spintronic Devices
Xiaobin Wang

MEMS: Fundamental Technology and Applications
Vikas Choudhary and Krzysztof Iniewski

Micro- and Nanoelectronics: Emerging Device Challenges and Solutions
Tomasz Brozek

Microfluidics and Nanotechnology: Biosensing to the Single Molecule Limit
Eric Lagally

MIMO Power Line Communications: Narrow and Broadband Standards,
EMC, and Advanced Processing
Lars Torsten Berger, Andreas Schwager, Pascal Pagani, and Daniel Schneider

PUBLISHED TITLES:

Mixed-Signal Circuits
Thomas Noulis

Mobile Point-of-Care Monitors and Diagnostic Device Design
Walter Karlen

Multisensor Data Fusion: From Algorithm and Architecture Design to Applications
Hassen Fourati

Nano-Semiconductors: Devices and Technology
Krzysztof Iniewski

Nanoelectronic Device Applications Handbook
James E. Morris and Krzysztof Iniewski

Nanomaterials: A Guide to Fabrication and Applications
Sivashankar Krishnamoorthy

Nanopatterning and Nanoscale Devices for Biological Applications
Šeila Selimović

Nanoplasmonics: Advanced Device Applications
James W. M. Chon and Krzysztof Iniewski

Nanoscale Semiconductor Memories: Technology and Applications
Santosh K. Kurinec and Krzysztof Iniewski

Novel Advances in Microsystems Technologies and Their Applications
Laurent A. Francis and Krzysztof Iniewski

Optical, Acoustic, Magnetic, and Mechanical Sensor Technologies
Krzysztof Iniewski

Optical Fiber Sensors: Advanced Techniques and Applications
Ginu Rajan

Optical Imaging Devices: New Technologies and Applications
Ajit Khosla and Dongsoo Kim

Organic Solar Cells: Materials, Devices, Interfaces, and Modeling
Qiquan Qiao

Physical Design for 3D Integrated Circuits
Aida Todri-Sanial and Chuan Seng Tan

Radiation Detectors for Medical Imaging
Jan S. Iwanczyk

Radiation Effects in Semiconductors
Krzysztof Iniewski

Reconfigurable Logic: Architecture, Tools, and Applications
Pierre-Emmanuel Gaillardon

Semiconductor Radiation Detection Systems
Krzysztof Iniewski

PUBLISHED TITLES:

Smart Grids: Clouds, Communications, Open Source, and Automation
David Bakken

Smart Sensors for Industrial Applications
Krzysztof Iniewski

Soft Errors: From Particles to Circuits
Jean-Luc Autran and Daniela Munteanu

Solid-State Radiation Detectors: Technology and Applications
Salah Awadalla

Technologies for Smart Sensors and Sensor Fusion
Kevin Yallup and Krzysztof Iniewski

Telecommunication Networks
Eugenio Iannone

Testing for Small-Delay Defects in Nanoscale CMOS Integrated Circuits
Sandeep K. Goel and Krishnendu Chakrabarty

VLSI: Circuits for Emerging Applications
Tomasz Wojcicki

Wireless Technologies: Circuits, Systems, and Devices
Krzysztof Iniewski

Wireless Transceiver Circuits: System Perspectives and Design Aspects
Woogeun Rhee

FORTHCOMING TITLES:

Advances in Imaging and Sensing
Shuo Tang, Dileepan Joseph, and Daryoosh Saeedkia

Analog Electronics for Radiation Detection
Renato Turchetta

Circuits and Systems for Security and Privacy
Farhana Sheikh and Leonel Sousa

Magnetic Sensors: Technologies and Applications
Kirill Poletkin

MRI: Physics, Image Reconstruction, and Analysis
Angshul Majumdar and Rabab Ward

Multisensor Attitude Estimation: Fundamental Concepts and Applications
Hassen Fourati and Djamel Eddine Chouaib Belkhiat

Nanoelectronics: Devices, Circuits, and Systems
Nikos Konofaos

Power Management Integrated Circuits and Technologies
Mona M. Hella and Patrick Mercier

FORTHCOMING TITLES:

Radio Frequency Integrated Circuit Design
Sebastian Magierowski

Semiconductor Devices in Harsh Conditions
Kirsten Weide-Zaage and Malgorzata Chrzanowska-Jeske

Smart eHealth and eCare Technologies Handbook
Sari Merilampi, Lars T. Berger, and Andrew Sirkka

Structural Health Monitoring of Composite Structures Using Fiber Optic Methods
Ginu Rajan and Gangadhara Prusty

Tunable RF Components and Circuits: Applications in Mobile Handsets
Jeffrey L. Hilbert

Wireless Medical Systems and Algorithms: Design and Applications
Pietro Salvo and Miguel Hernandez-Silveira

Analog Electronics for Radiation Detection

Edited by
Renato Turchetta
Science and Technology Facilities Council (STFC), United Kingdom

Krzysztof Iniewski
MANAGING EDITOR
Emerging Technologies CMOS Services Inc.
Vancouver, British Columbia, Canada

CRC Press
Taylor & Francis Group
Boca Raton London New York

CRC Press is an imprint of the
Taylor & Francis Group, an **informa** business

CRC Press
Taylor & Francis Group
6000 Broken Sound Parkway NW, Suite 300
Boca Raton, FL 33487-2742

First issued in paperback 2018

© 2016 by Taylor & Francis Group, LLC
CRC Press is an imprint of Taylor & Francis Group, an Informa business

No claim to original U.S. Government works

ISBN-13: 978-1-4987-0356-7 (hbk)
ISBN-13: 978-1-138-58602-4 (pbk)

Library of Congress Cataloging-in-Publication Data

Names: Turchetta, Renato, editor.
Title: Analog electronics for radiation detection / editor, Renato Turchetta.
Description: Boca Raton : Taylor & Francis, CRC Press, 2016. | Series:
Devices, circuits, and systems ; 59 | Includes bibliographical references
and index.
Identifiers: LCCN 2015047406 | ISBN 9781498703567 (alk. paper)
Subjects: LCSH: Nuclear counters.--Design and construction. | Analog
electronic systems. | Analog integrated circutits. |
Radioactivity--Measurement.
Classification: LCC TK9180 .A53 2016 | DDC 539.7/7--dc23
LC record available at http://lccn.loc.gov/2015047406

Visit the Taylor & Francis Web site at
http://www.taylorandfrancis.com

and the CRC Press Web site at
http://www.crcpress.com

Contents

Preface..xi
Editors... xiii
Contributors xv

Chapter 1   Integrated Analog Signal Processing Readout Front Ends
for Particle Detectors... 1

Thomas Noulis

Chapter 2   Analog Electronics for HVCMOS Sensors 27

Ivan Peric

Chapter 3   Analog Electronics for Radiation Detection 47

Juan A. Leñero-Bardallo and Ángel Rodríguez-Vázquez

Chapter 4   Low-Noise Detectors through Incremental Sigma–Delta ADCs....... 71

Adi Xhakoni and Georges Gielen

Chapter 5   Time-to-Digital Conversion 91

Yasuo Arai

Chapter 6   Digital Pulse-Processing Techniques for X-Ray
and Gamma-Ray Semiconductor Detectors..................................... 121

Leonardo Abbene, Gaetano Gerardi, and Fabio Principato

Chapter 7   Silicon Photomultipliers for High-Performance Scintillation
Crystal Readout Applications.. 141

Carl Jackson, Kevin O'Neill, Liam Wall, and Brian McGarvey

Chapter 8   Designing Photon-Counting, Wide-Spectrum Optical Radiation
Detectors in CMOS-Compatible Technologies................................ 185

Edoardo Charbon and Chockalingam Veerappan

Chapter 9 Front-End Electronics for Silicon Photomultipliers 203

Cristoforo Marzocca, Fabio Ciciriello, Francesco Corsi,
Francesco Licciulli, and Gianvito Matarrese

Chapter 10 CMOS Image Sensors for Radiation Detection 237

Nicola Guerrini

Chapter 11 Technology Needs for Modular Pixel Detectors 265

Paul Seller

Index 283

Preface

We live in an increasingly digital world. Streams of bits form the layer that conveys the information that keeps us connected together at a global scale that was never experienced before. We are getting so used to dealing with information through digital interfaces that it can be easy to forget that the world surrounding us is analog. If we want to know the weather forecast where we live, or where we are traveling to, we query a digital application, but the forecast will be based on the digitized data that are delivered by analog sensors, measuring parameters like temperature, pressure, and so on. Digital data associated with each single pixel deliver images, but, behind that, there is an image sensor, where analog data are converted into digital numbers. The latter is an example where the information about our surroundings is brought to us by radiation, in this case, visible light photons. This is by far the most popular example of radiation detection, intended as photons or particles that can either directly or indirectly generate electric signals in a suitable medium. But types of radiation other than visible light also surround us in nature or are purposely generated as they can provide useful, complementary information: x-rays are used for medical diagnostics or luggage scans; gamma rays are used for medical applications; and particles are accelerated and studied to understand the structure of matter. Different types of detectors are needed for different applications and to transform the charge that is generated by radiation into a stream of bits that can be used by our computers; analog electronics are used. Analog electronics will process the charge that is generated in the detectors and extract useful information. Whenever individual quantum of radiation impinges too fast over the detector, the signal-conditioning electronics will integrate the charge and transform it into a voltage that is proportional to the amount of detected charge. If the electronics are fast enough to deal with each individual quantum separately, more detailed information can be extracted, like the energy or wavelength of the quantum, its arrival time, the duration of a pulse, its shape, and so on.

The aim of this book is to cover a broad range of analog circuitry that is employed to extract this information from the impinging radiation. Thomas Noulis reviews the main design parameters of front-end circuitry that is developed in a microelectronics technology. The introduction of very-large-scale integration (VLSI) technology for the readout of radiation detectors became a necessity with their ever-increasing segmentation. Solid-state microstrip detectors were introduced in the early 1980s and, soon after, dedicated VLSI circuitry was developed. Parallel to this development, pixelated detectors were being developed for the infrared focal plane as well as for integrating x-ray sensors. At the end of the same decade and thanks to the increasing miniaturization brought by microelectronics scaling, typical radiation-detector signal-conditioning electronics were moved inside the pixel, triggering the rise of hybrid pixel detectors for particle physics. The charge signal generated by a particle is relatively small, and careful optimization of the noise performance of the front-end electronics is required. Thomas Noulis's chapter covers these aspects, giving the readers guidelines for understanding and modeling the front end.

Hybrid pixel detectors, where the readout integrated circuit (ROIC) is connected, typically by bump bonding, to a separately optimized detector layer, are suitable for applications like particle physics, where expensive, highly optimized detectors can be used. In the 1990s, complementary metal–oxide semiconductor (CMOS) image sensors, where a single chip integrates the readout electronics and the sensing element in a single piece of silicon, were invented. Nowadays, we are all used to having them in our pockets, with most consumer imaging products being based on image sensors. A few years after their invention, CMOS image sensors were proposed for the detection of charged particles and are now routinely used for transmission electron microscopy and are starting to be used in particle physics as well. Nicola Guerrini's chapter introduces the basis for the use of CMOS image sensors for the detection of charged particles and other nonconsumer applications. Future use of these sensors in radiation-harsh environments like the Large Hadron Collider at European Organization for Nuclear Research (CERN) is the focus of the chapter by Ivan Peric. He introduces the use of a high-voltage CMOS process to provide a higher bias voltage to the sensing part.

Moving from the conditioning of the analog signal into the digital domain is the subject of a few chapters of this book. Juan A. Leñero-Bardallo and Ángel Rodríguez-Vázquez present an overview of analog-to-digital converters (ADCs), discussing the pros and cons of ADCs that are integrated at pixel level, column level, or per chip. Incremental sigma–delta ADCs are a relatively new architecture that could provide speed in a compact converter. This is discussed separately in the chapter by Adi Xhakoni and Georges Gielen. As mentioned above, it is sometimes necessary to digitize quantities other than amplitude. In his chapter, Yasuo Arai discusses the different architecture of time-to-digital converters, a circuit block that is used to convert time information, like the interval between two events, into a digital number. Digital pulse-processing techniques are discussed in the chapter by Leonardo Abbene and his coauthors.

In many applications, there is a need to detect a very small amount of visible light photons. In this case, the front-end electronics need to be able to generate a large voltage signal even for a single detected photon. Carl Jackson and his coauthors discuss the fundamental parameters that are associated to a silicon photomultiplier, a device used for single visible-light photon detection. Silicon photomultipliers are also at the core of the chapter by Cristoforo Marzocca and his collaborators. In this chapter, the different types of front end are discussed. Details of circuit solutions are provided together with examples of integration of these architectures in VLSI circuits. In recent years, single photon detectors have also been integrated directly in CMOS. Continuous improvements have enabled good-quality detectors in this technology. High integration of functionalities is thus achieved, and pixel sensors with per-pixel time-to-digital converters can, for example, be found. The chapter from Edoardo Charbon and Chockalingam Veerappan covers this interesting topic.

Taking the signal out of the detector and into the electronics can be a challenge, especially as the density of channels increases. The chapter from Paul Seller covers this aspect with a look also at emerging 3D technology, where each layer of the electronics can be optimized separately to provide optimized performance in a compact device.

Renato Turchetta

Editors

Renato Turchetta is leading the development of high-end complementary metal–oxide semiconductor (CMOS) image sensors at the Rutherford Appleton Laboratory in Didcot, UK, the largest national laboratory in the United Kingdom. He earned his master's degree (Laurea) from the University of Milan, Italy, in 1988 and his PhD in applied physics from the University of Strasbourg, France, in 1991. He worked as an assistant professor there until 1999, the year when he moved to the Rutherford Appleton Laboratory. In his career, he has worked on the development of semiconductor detectors and their readout microelectronics before focusing on CMOS image sensors. He has worked in this area for over 15 years. He is the author and coauthor of several patents and over 100 papers that were published in international journals. He is currently an acting member of the scientific committee for Image Sensors Europe, the International Congress on High-Speed Imaging and Photonics, and the Pixel and Workshop on CMOS Active Pixel Sensors for Particle Tracking (CPIX) conferences. Since 2015, he is also the director of vivaMOS Ltd, a high-tech company that specializes in the fabrication of large-area CMOS image sensors for x-ray detection.

Krzysztof (Kris) Iniewski is managing research and development at Redlen Technologies Inc., a startup company in Vancouver, Canada. Redlen's revolutionary production process for advanced semiconductor materials enables a new generation of more-accurate, all-digital, radiation-based imaging solutions. Kris is also a founder of Emerging Technologies CMOS Inc. (ET CMOS Inc.) (http://www.etcmos.com), an organization of high-tech events covering communications, microsystems, optoelectronics, and sensors. In his career, Dr. Iniewski has held numerous faculty and management positions at the University of Toronto, University of Alberta, Simon Fraser University (SFU), and PMC-Sierra Inc. He has published over 100 research papers in international journals and conferences. He holds 18 international patents that were granted in the United States, Canada, France, Germany, and Japan. He is a frequently invited speaker and has consulted for multiple organizations internationally. He has written and edited several books for CRC Press, Cambridge University Press, IEEE Press, Wiley, McGraw-Hill, Artech House, and Springer. His personal goal is to contribute to healthy living and sustainability through innovative engineering solutions. In his leisure time, Kris can be found hiking, sailing, skiing, or biking in beautiful British Columbia. He can be reached at kris.iniewski@gmail.com.

Contributors

Leonardo Abbene
Dipartimento di Fisica e Chimica
University of Palermo
Palermo, Italy

Yasuo Arai
High Energy Accelerator Research
    Organization (KEK)
Institute of Particle and Nuclear Studies
Tsukuba, Ibaraki, Japan

Edoardo Charbon
Department of Electrical Engineering,
    Mathematics, and Computer
    Sciences
Delft University of Technology
Delft, The Netherlands

Fabio Ciciriello
Department of Electrical and
    Information Engineering
Politecnico di Bari
Bari, Italy

Francesco Corsi
Department of Electrical and
    Information Engineering
Politecnico di Bari
Bari, Italy

Gaetano Gerardi
Dipartimento di Fisica e Chimica
University of Palermo
Palermo, Italy

Georges Gielen
Department of Electrical Engineering
KU Leuven
Leuven, Belgium

Nicola Guerrini
Science and Technology Facilities
    Council (STFC)
Rutherford Appleton Laboratory
Didcot, United Kingdom

Carl Jackson
SensL Technologies Ltd.
Cork, Ireland

Juan A. Leñero-Bardallo
University of Sevilla
Seville, Spain

Francesco Licciulli
Department of Electrical and
    Information Engineering
Politecnico di Bari
Bari, Italy

Cristoforo Marzocca
Department of Electrical and
    Information Engineering
Politecnico di Bari
Bari, Italy

Gianvito Matarrese
Department of Electrical and
    Information Engineering
Politecnico di Bari
Bari, Italy

Brian McGarvey
SensL Technologies Ltd.
Cork, Ireland

Thomas Noulis
Electronics Lab. of the Physics
    Department
Aristotle University of Thessaloniki
Thessaloniki, Greece

Kevin O'Neill
SensL Technologies Ltd.
Cork, Ireland

Ivan Peric
Ruprecht-Karls-Universität Heidelberg
Heidelberg, Germany

Fabio Principato
Dipartimento di Fisica e Chimica
University of Palermo
Palermo, Italy

Ángel Rodríguez-Vázquez
University of Sevilla
Seville, Spain

Paul Seller
Rutherford Appleton Laboratory
Didcot, United Kingdom

Chockalingam Veerappan
Department of Electrical Engineering,
    Mathematics, and Computer
    Sciences
Delft University of Technology
Delft, The Netherlands

Liam Wall
SensL Technologies Ltd.
Cork, Ireland

Adi Xhakoni
Department of Electrical Engineering
KU Leuven
Leuven, Belgium

1 Integrated Analog Signal Processing Readout Front Ends for Particle Detectors

Thomas Noulis

CONTENTS

1.1 Introduction 1
1.2 Readout Front-End Analog Processing Channel 2
    1.2.1 Preamplifier–Shaper Structure 2
    1.2.2 CSP–Shaping Filter System Noise Analysis and Optimization 2
1.3 CSP–Shaper System Design 6
    1.3.1 CSP Design 6
1.4 S-G Shaping Filter Design 10
1.5 Operational Transconductance Amplifier–Based Shaper Approach 13
1.6 Advanced CSP–Shaper Analog Signal Processor Implementation 17
1.7 Experimental Results 20
1.8 Current Challenges and Limitations 23
References 24

1.1 INTRODUCTION

The current trend in high-energy physics, biomedical applications, radioactivity control, space science, and other disciplines that require radiation detectors is toward smaller, higher-density systems to provide better position resolution. Miniaturization, low power dissipation, and low-noise performance are stringent requests in modern instrumentation where portability and constant increase of channel numbers are the main streamlines. In most cases, complementary metal–oxide semiconductor (CMOS) technologies have fully proven their adequacy for implementing data acquisition architectures based on functional blocks such as charge preamplifiers, continuous time or switch–capacitor filters, sample-and-hold amplifiers, analog-to-digital converters, etc., in analog signal processing for particle physics, nuclear physics, and x- or beta-ray detection [1–3].

Several motivations suggest that most of these applications can benefit from the use of application-specific integrated circuit (ASIC) readouts instead of discrete solutions. The most crucial motivation is that the implementation of readout electronics and semiconductor detectors onto the same die offers enhanced detection sensitivity, thanks to improved-noise performances [4–8]. Placing the very first stage of the front-end circuit close to the detector electrode reduces the amount of material and complexity in the active detection area and minimizes connection-related stray capacitances. This method allows the noise optimization theory predictions [9,10] to be satisfied effectively, especially in the case of silicon sensors with very low anode capacitance such as silicon drift detectors, semiconductor charge-coupled devices, and pixels. On the other hand, the use of discrete transistors, with their relatively high (a few picofarads) gate capacitances, as front-end elements of hybrid circuits, cannot comply with the stringent low-noise requirements. As a result, continuous efforts were performed in order to implement readout systems in monolithic form, and CMOS and silicon germanium (SiGe) BiCMOS technologies have been chosen due to their high integration density, relatively low power consumption, and capability to combine analog and digital circuits on the same chip.

1.2 READOUT FRONT-END ANALOG PROCESSING CHANNEL

1.2.1 PREAMPLIFIER–SHAPER STRUCTURE

The preamplifier–shaper structure is commonly adopted in the design of the radiation detection front end systems. A block diagram of such a detection system is shown in Figure 1.1. An inverse-biased diode (Si or Ge) detects radiation events, generating electron–hole pairs that are proportional to the absorbed energies. A low-noise charge-sensitive preamplifier (CSP) is used at the front end due to its low-noise configuration and the insensitivity of the gain to the detector capacitance. The generated charge Q is integrated onto a small feedback capacitance, which gives rise to a step voltage signal at the output of the CSP with an amplitude that is equal to Q/C_f. This is fed to a main amplifier, called a pulse shaper, where pulse shaping is performed to optimize the signal/noise ratio (SNR). The resulting output signal is a narrow pulse that is suitable for further processing.

1.2.2 CSP–SHAPING FILTER SYSTEM NOISE ANALYSIS AND OPTIMIZATION

In order to analyze clearly the noise-matching mechanism in monolithic implementations, it is necessary to briefly review the noise characteristics of the

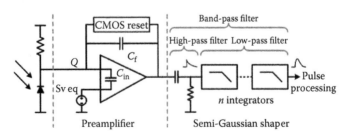

FIGURE 1.1 Preamplifier–shaping filter readout front-end system. Sv eq, equivalent input noise.

metal–oxide-semiconductor field-effect transistor (MOSFET). In CMOS technology, two major noise sources exist: (1) thermal noise and (2) flicker noise (or $1/f$ noise). Using a basic metal–oxide semiconductor (MOS) transistor model and the Nyquist theory, the MOS drain current thermal noise spectral density, in the saturation regime, is given by Refs. [9,11,12]

$$i_d^2(f) = \frac{8}{3} kT\mu C_{ox}\left(\frac{W}{L}\right)(V_{GS} - V_T) = \frac{8}{3} kTg_m \qquad (1.1)$$

where k is the Boltzmann constant; T is the absolute temperature; μ is the carrier mobility; g_m is the MOSFET transconductance; and variables W, L, and C_{ox} represent the transistor's width, length, and gate capacitance per unit area, respectively. In contrast to the channel thermal noise, which is well understood, the flicker noise mechanism is more complex, although its presence is surprisingly universal in all types of semiconductor devices [13–16]. A large number of theoretical and experimental studies show that the $1/f$ noise in MOSFET is caused by the random trapping and detrapping of the mobile carriers in the traps that are located at Si-SiO$_2$ interface and within the gate oxide. On the basis of this model, the short-circuit drain current noise spectral density in the saturation region is given by Refs. [9,12]

$$i_f^2(f) = \frac{K_F I_{DS}}{C_{ox} L^2 f} \qquad (1.2)$$

where K_F is a technologically dependent constant that is proportional to the effective trap density $N_t(F_n)$, and I_{DS} is the drain current. Dividing Equation 1.2 by the square of the transconductance, the equivalent input $1/f$ noise can easily be calculated:

$$v_f^2(f) = \frac{K_F}{2\mu C_{ox}^2 WLf} = \frac{K_f}{C_{ox}^2 WLf} \qquad (1.3)$$

The $1/f$ noise voltage depends only on the gate area, and its amplitude is dependent on the $1/f$ noise coefficient, which value is variable in relation to the V_{gs} voltage of the MOS transistor [14,17].

In addition to the channel thermal and flicker noise, MOS transistors exhibit parasitic noise due to the resistive polygate and substrate resistance. These parasitic noise sources in modern semiconductor technologies should be taken into account since the polygate resistance R_g and the substrate resistance R_b depend mainly on the layout structure of MOS transistors, and the related modeling is quite advanced—also, by using specific layout techniques, their noise contribution can be minimized. However, these noise sources are greatly critical in high-frequency applications and, in the case of the detector readout integrated circuits (ICs), can be considered as negligible.

The respective equivalent simplified input noise MOSFET model is shown in Figure 1.2 [18].

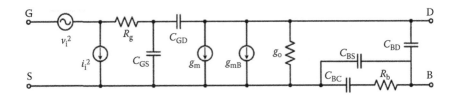

FIGURE 1.2 Equivalent input noise generator MOS small-signal model.

This model is based on the fact that the noise performance of any two-port network can be represented by two equivalent noise generators at the input stage of the network [19]. The two equivalent input noise generators are calculated from Figure 1.2, and they are given in Equations 1.4 and 1.5:

$$v_i^2(f) = \frac{\left(i_d^2 + i_f^2\right)}{\left|g_m - j\omega C_{GD}\right|^2} \tag{1.4}$$

$$i_i^2(f) = \left|j\omega\left(C_{GS} + C_{GD}\right)\right|^2 \frac{\left(i_d^2 + i_f^2\right)}{\left|g_m - j\omega C_{GD}\right|^2} \tag{1.5}$$

Since the factor $\left(g_m/2\pi C_{GD}\right)$ is much higher than the transistor cutoff frequency f_T, the term $j\omega C_{GD}$ can be neglected with respect to g_m for all the practical cases of interest. It is also essential to note that these two terms of Equations 1.4 and 1.5 are fully correlated.

High-density-semiconductor front-end systems are designed according to low-noise criteria. The noise performance of a detector readout system is generally expressed as the *equivalent noise charge* (*ENC*) and is defined as the ratio of the total *root mean square* (*RMS*) noise at the output of the pulse shaper to the signal amplitude due to one electron charge q. The noise contribution of the amplification stage is the dominant source that determines the overall system noise and should therefore be optimized. The main noise contributor of the analog part is the CSP input MOSFET, and the noise types associated with this device are 1/f and channel thermal noise [12].

The total noise power spectrum at the CSP output due to the thermal and flicker noise sources is given by Refs. [9,10]

$$v_0^2(j\omega) = \left|\frac{C_t}{C_f}\right|^2 v_{eq}^2 = \left|\frac{C_t}{C_f}\right|^2 \left(\frac{8}{3} kT \frac{1}{g_m} + \frac{K_f}{C_{ox}^2 WLf}\right)^2 \tag{1.6}$$

where C_t is the total CSP input capacitance, and υ_{eq} is the total equivalent noise voltage of the input MOS transistor. The total integrated RMS noise is given in the following equation:

$$\upsilon_{total}^2 = \int_0^\infty |\upsilon_0(j\omega)|^2 |H(j\omega)|^2 \, df \tag{1.7}$$

where $H(j\omega)$ is the transfer function consisting of a semi-Gaussian (S-G) pulse shaper. The ENCs due to thermal and $1/f$ noise, considering the expressions 1.4 through 1.7 and for an amplifier with capacitive source, are given by Refs. [9,10]

$$\mathrm{ENC}_{th}^2 = \frac{8}{3} kT \frac{1}{g_m} \frac{C_t^2 B\left(\frac{3}{2}, n-\frac{1}{2}\right) n}{q^2 4\pi\tau_s} \left(\frac{n!^2 \, e^{2n}}{n^{2n}}\right) \tag{1.8}$$

$$\mathrm{ENC}_{1/f}^2 = \frac{K_f}{C_{ox}^2 \, WL} \frac{C_t^2}{q^2 2n} \left(\frac{n!^2 \, e^{2n}}{n^{2n}}\right) \tag{1.9}$$

where B is the beta function [9], q is the electronic charge, τ_s is the peaking time of the shaper, and n is the order of the S-G shaper.

Capacitances C_d, C_f, C_{GS}, and C_{GD} are the capacitances of the detector, feedback, gate–source, and gate–drain of the input MOSFET, respectively, and C_p is the parasitic capacitance of the interconnection between the detector and the amplifier input, which in this application is considered as negligible. The total input stage capacitance is given by Refs. [9,18,20,21]

$$C_t = C_{totalin} = C_d + C_p + C_f + C_{GS} + C_{GD} \tag{1.10}$$

Input MOSFET optimum gate widths exist for which the respective thermal and flicker ENCs are minimal. These optimum dimensions are extracted by minimizing the respective ENCs of Equations 1.8 and 1.9, respectively [9]. The optimal gate widths are given in the following equation:

$$\frac{\theta \mathrm{ENC}_{th}}{\theta W} = 0 \Rightarrow W_{th} = \frac{C_d + C_f}{2C_{ox}\alpha L} \tag{1.11}$$

$$\frac{\theta \mathrm{ENC}_{1/f}}{\theta W} = 0 \Rightarrow W_{1/f} = 3W_{th} \tag{1.12}$$

where α is defined as $\alpha = 1 + \dfrac{9X_j}{4L}$, and X_j is the metallurgical junction depth. Equations 1.11 and 1.12 are valid when capacitance C_d is in the range of picofarads.

It can easily be determined which noise component (thermal or $1/f$) dominates by calculating if the ratio of the respective ENCs is beyond or below unity. The noise comparison ratio (NCR) also depends on the technology, comes from Equations 1.6 and 1.7, and is given by Ref. [20]

$$NCR^2 = \frac{ENC_{th}^2}{ENC_{1/f}^2} = \frac{8kT2n^2B\left(\dfrac{3}{2}, n-\dfrac{1}{2}\right)C_{ox}^2 L^{3/2}}{3\sqrt{2\mu C_{ox}} \cdot 4\pi K_f} \cdot \frac{1}{\tau_s \sqrt{I_{DS}}}\sqrt{W} \qquad (1.13)$$

When it is known, for a given technology and for a MOSFET operating in the saturation regime, which noise component, thermal or $1/f$, dominates, the type of the preamplifier input MOSFET and its optimum dimensions can be selected, considering that, typically, p-channel transistors have less $1/f$ noise than their n-channel counterparts, since their majority carriers (holes) are less likely to be trapped due to their lower mobility [18].

As it is obvious, the CSP input noise is in practice defined by the detector capacitance, the peaking time specification, and the selection of the input MOS type and dimensions (as it was analyzed in Section 1.2.2). A large detector capacitance results in a higher input noise, as it affects the input node biasing. Additionally, the thermal and flicker equivalent noise voltages (ENCs) are proportional to C_t, which values are mainly affected by C_d. The peaking time specification determines the operating bandwidth of the readout system. Consequently, it determines the flicker or thermal noise dominance since the $1/f$ noise is negligible in the frequency region after 10 kHz. Concerning Equations 1.11 and 1.12, these formulas are the result of the optimization methodology, and they are used in calculating the NCR (Equation 1.13). The dimensions of the input MOSFET are also selected according to these equations ($W_{1/f}$ if it is a PMOS or W_{th} if it is an NMOS) [18].

1.3   CSP–SHAPER SYSTEM DESIGN

1.3.1   CSP DESIGN

The CSP (Figure 1.3) is commonly implemented using a folded-cascode structure, built of transistors M1, M2, and M4. This architecture provides a high direct current (DC) gain and a relatively large operating bandwidth. The folded-cascode amplifier has an n-channel input transistor (or drive), whereas a p-channel transistor is used for the cascode (or common-gate) transistor. A complementary structure can also be configured always in relation to the noise specifications of the application. This configuration allows the DC level of the output signal to be the same as the DC level of the input signal [22,23]. In case the circuit is designed to operate with DC coupling detectors, the CSP should be able to supply a current in the range of a few picoamperes to a few nanoamperes through the feedback loop in order to match the detector leakage current. The open-loop gain of such configuration is determined by the ratio of currents in the left and right branch of the folded cascode, the sizes of the input transistor, the load transistor, and the total current I_{d1} in the input transistor.

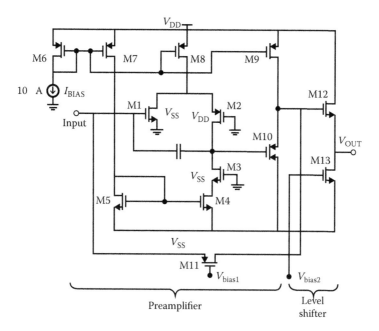

FIGURE 1.3 CSP using a folded-cascode amplifier.

Regarding the CSP reset mechanism, the integrated reset device provides a continuous discharge of the feedback capacitance, a DC path for the detector leakage current, and defines the DC operating point of the amplifier. A simple resistor would not be a feasible reset device. In fact, in order to make its thermal noise negligible compared with the shot noise that is associated with the leakage current, its value should not be relatively high (e.g., a 500-MΩ resistor introduces an amount of thermal noise that is equal to the shot noise of 100-pA leakage current) and cannot be integrated on a small silicon area. An active device (MOS or bipolar) should be used instead [24]. The CSP reset device was configured with a PMOS transistor (M11) that is biased in the triode region in order to avoid the use of a high-value resistance (Figure 1.3).

Transistors M9 and M10 implement an output buffer for the CSP circuit. The current mirror architecture of transistors M6, M7, and M8 are biasing using the I_{bias} current source. The feedback capacitance C_f is placed between the input node and the gate of the source follower stage to avoid introduction on the closed loop of the follower stage complex poles and to isolate the C_f from the following stage. This feedback capacitance C_f is discharged via the input node, which is connected to both C_f and transistor M11. Transistor M3 is in cascade configuration with M4, the MOSFET comprising the current mirror M5, and it was placed in the CSP in order to have a higher output resistance.

In the specific design, the maximum signal swing must be limited for keeping all the devices in the saturation regime, i.e., $V_{GS} > V_T$ and $V_{DS} > V_{GS} - V_T$. Therefore, the bias current source I_{bias} has been selected with a specific low value in order to achieve a low power dissipation performance.

The simplified small-signal equivalent circuit of the core folded-cascode amplifier without the feedback capacitance, the reset device, and transistor M3, which is in cascade configuration with M4, is shown in Figure 1.4.

The open-loop gain $A_V = (V_0/V_i)$ is given by

$$A_V = \frac{-\dfrac{g_{m1}}{g_{ds4}}\left(1 - s\dfrac{C_{gd1}}{g_{m1}}\right)}{\left[1 + sC_{ds1}R_1\left(1 + \dfrac{g_{m1}}{g_{m2}}\right)\right]\left[1 + s\dfrac{C_{p1} + C_{gd1}}{g_{m2}}\right]\left(1 + s\dfrac{C_{p2}}{g_{ds4}}\right)} \tag{1.14}$$

where

$$C_{p1} = C_{gd1} \tag{1.15}$$

$$C_{p2} = C_{ds1} + C_{gd8} + C_{ds8} + C_{gs2} \tag{1.16}$$

$$C_{p3} = C_{gd2} + C_{gd4} + C_{bd2} + C_{bd4} \tag{1.17}$$

$$R_1 = \frac{1}{g_{ds1} + g_{ds8}} \tag{1.18}$$

$$R_2 = \frac{1}{g_{ds4}} \tag{1.19}$$

The DC gain A_{V0} of the two-port network is given in the following equation:

$$A_{V0} = \frac{-g_{m1}}{g_{ds4}} = -g_{m1}R_2 \tag{1.20}$$

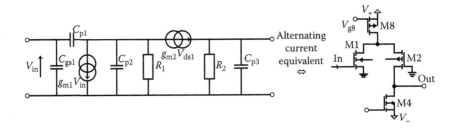

FIGURE 1.4 Small-signal equivalent circuit of the open-loop preamplifier.

and the dominant pole is

$$P_1 = \frac{-g_{ds4}}{C_{p3}} = -\frac{1}{C_{p3}R_2} \tag{1.21}$$

The P-MOSFET-based equivalent resistor operating as reset device has a value of

$$R_{eq} = \sqrt{\frac{1}{2I\mu C_{ox}(W/L)}} \tag{1.22}$$

where I is the drain current.

Using the specific reset device implementation, the equivalent resistor can be externally modified by fixing suitably the gate bias voltage of the P-MOSFET-based equivalent resistor. However, since the gate bias voltage is conditioned by the other parameters (e.g., the power consumption specification, the process and its allowable supply voltages, the charge–discharge time specification), the dimensions of the feedback MOS are also suitably selected according to each application [18].

Considering the noise analysis of the folded-cascode structure, shown in Figure 1.5, the basic structure of the folded-cascode amplifier (in relation to our preamplifier configuration) is depicted including all the equivalent MOSFET noise current sources.

The total noise current is equal to

$$I_{o(sc)} = g_{m2}\left(-g_{m1}V_i - I_{n1} - I_{n2} + I_{n8} + g_{m8}(I_{nb} - I_{n6})\left(r_{06}//g_{m6}^{-1}\right)\right)$$

$$\cdot\left(r_{01}//r_{08}//r_{02}//g_{m2}^{-1}\right) + I_{n2} \cong -g_{m1}V_i - I_{n1} + I_{n8} + g_{m8}\left(\frac{I_{nb} - I_{n6}}{g_{m6}}\right) \tag{1.23}$$

$$\cong -g_{m1}(V_i + V_{ni})$$

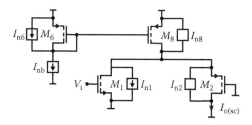

FIGURE 1.5 Folded-cascode amplifier equivalent schema structure with noise current sources.

Performing the approximations $\left(r_{06}//g_{m6}^{-1}\right) \cong g_{m6}^{-1}$ and $\left(r_{01}//r_{08}//r_{02}//g_{m2}^{-1}\right) \cong g_{m2}^{-1}$, the equivalent noise voltage is given by

$$V_{ni} = g_{m1}\left[-I_{n1} + I_{n8} + \frac{g_{m8}}{g_{m6}}(I_{nb} - I_{n6})\right] \tag{1.24}$$

The respective square value is

$$v_{ni}^2 = g_{m1}^2\left[i_{n1}^2 + i_{n8}^2 + \left(\frac{g_{m8}}{g_{m6}}\right)^2 \cdot \left(i_{nb}^2 + i_{n6}^2\right)\right] \tag{1.25}$$

While, theoretically, there is noise contribution generated by M6, M8, and the DC bias current source, practically, their contribution can be neglected when reflected to the amplifier topology input. As it is shown by Equations 1.21 through 1.25, the generated noise current of each MOS transistor is separately dependent upon its drain current. Specifically, these equations show that low-noise performance in a MOSFET requires a large value of transconductance g_m, which, in turn, means that the transistor should have a large width/length (W/L) ratio and be operated in a large, quiescent current level. Practically, as the drain current gets higher, the noise performance decreases, and the power consumption increases. As it is obvious from the preamplifier circuit, the current flowing through M8 (which is practically obtained from the current source and the respective current mirror structure) is *shared* in M1 and M2 of the folded-cascode amplifier. Taking in consideration the large dimensions of M1, its noise contribution is the dominant noise source of the CSP, and, therefore, the noise methodology is focused on its noise optimization. Hence, the full noise optimization methodology is focused on the input node conditions and consequently on the input transistor type and dimensions. Other design parameters that can be optimized in terms of noise performance are the peaking time and the order of the shaper, always in relation to the radiation-detection application specifications.

1.4   S-G SHAPING FILTER DESIGN

S-G pulse-shaping filters are the most common pulse shapers that are employed in readout systems; their use in electronics spectrometer instruments is to measure the energy of charge particles [25], and their purpose is to provide a voltage pulse whose height is proportional to the energy of the detected particle. The theory behind pulse-shaping systems, as well as different realization schemes, can be found in the literature [25,26]. It has been proven that a Gaussian-shaped step response provides optimum signal-to-noise characteristics. However, the ideal S-G shaper is noncasual and cannot be implemented in a physical system. A well-known technique to approximate a delayed Gaussian waveform is to use a CR-(RC)^n filter [25]. Such shaper principal schema is shown in Figure 1.6.

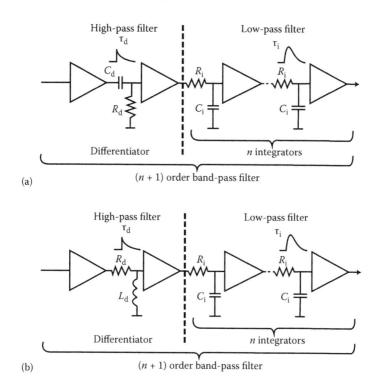

FIGURE 1.6 Principal diagram of a CR-(RC)^n shaping filter (a) with RC high-pass filter and (b) with RL high-pass filter.

A high-pass filter (HPF) sets the duration of the pulse by introducing a decay time constant. A low-pass filter (LPF), which follows, increases the rise time to limit the noise bandwidth. Although pulse shapers are often more sophisticated and complicated, the CR-(RC)^n shaper contains the essential features of all pulse shapers, a lower and an upper frequency bound, and it is basically an $(n + 1)$ order band-pass filter (BPF), where n is the integrator number. The transfer function of an S-G pulse shaper consisting of one CR differentiator and n RC integrators (Figure 1.6) is given by Refs. [9,10]

$$H(s)_{shaper} = \left(\frac{s\tau_d}{1 + s\tau_d} \right)\left(\frac{A}{1 + s\tau_i} \right)^n = H(s)_{BPF} \tag{1.26}$$

where τ_d is the time constant of the differentiator, τ_i is the time constant of the integrators, and A is the integrators' DC gain. The number n of the integrators is called shaper order. Peaking time is the time that the shaper output's signal reaches the peak amplitude and is defined by

$$\tau_s = n\tau_i = nRC \tag{1.27}$$

Increasing the value of n results in a step response that is closer to an ideal Gaussian but with larger delay. The order n and the peaking time τ_s, depending on the application, can or cannot be predefined by the design specifications.

The operating bandwidth of an S-G shaper is given by

$$\mathrm{BW} = f_i - f_d = \frac{1}{2\pi\tau_i} - \frac{1}{2\pi\tau_d} \tag{1.28}$$

The above combined band-pass frequency behavior enhances the SNR by separating the *noise sea* from the signal main Fourier components. The choice of the cutoff frequencies defines the shaper noise performance and consequently the SNR. In addition, the frequency domain transfer characteristics are strictly linked to the shaper time domain behavior or output pulse shape [20]. Pulse shaping has two conflicting objectives. The first is to restrict the bandwidth to match the measurement time. The second is to constrain the pulse so that successive signal pulses can be measured without undershoot or overlap (pileup). Reducing the pulse duration increases the allowable signal rate but at the expense of electronic noise. In designing the shaper, these conflicting goals should be balanced. Many different considerations lead to the *nontextbook* compromise that optimum shaping depends on the application [27].

Concerning the shaper peaking time, in order to achieve a predefined peaking time value from the application specification value, the shaper-model passive elements should be suitably selected.

The total CSP–second-order S-G shaper system transfer function using a Laplace representation is

$$H(s)_{\text{total}} = \left(\frac{A_{\text{pr}}}{1 + s\tau_{\text{pr}}}\right)\left(\frac{s\tau_d}{1 + s\tau_d}\right)\left(\frac{A_{\text{sh}}}{1 + s\tau_i}\right)^2 \tag{1.29}$$

where A_{pr} is the preamplifier gain, and τ_{pr} is its time rise constant. Considering a Dirac pulse $\delta(t)$ as an input signal, by taking the inverse Laplace transform of the product, the output signal in the time domain is given by Refs. [21,28,29]

$$h_{\text{total}}(t) = \int_{\sigma - j\omega}^{\sigma + j\omega} H_{\text{total}}(s) \cdot e^{st}\, ds \tag{1.30}$$

Solving the above integral, the output signal of the readout system is

$$h_{\text{total}}(t) = A_{\text{pr}} A_{\text{sh}} \cdot \left(k_1 e^{-t/\tau_{\text{pr}}} + k_2 e^{-t/\tau_d} + k_3 e^{-t/\tau_i} + k_4 t e^{-t/\tau_i}\right) \tag{1.31}$$

where k_1, k_2, k_3, and k_4 are the following constants [21,28,29]:

$$k_1 = \frac{\dfrac{1}{\tau_{pr}}}{\left(\dfrac{1}{\tau_{pr}}\right)^3 - \left(\dfrac{1}{\tau_{pr}}\right)^2 \left(\dfrac{1}{\tau_d} + 2\dfrac{1}{\tau_i}\right) + \dfrac{1}{\tau_{pr}} \cdot \dfrac{1}{\tau_i}\left(\dfrac{1}{\tau_i} + 2\dfrac{1}{\tau_d}\right) - \dfrac{1}{\tau_d}\left(\dfrac{1}{\tau_i}\right)^2} \tag{1.32}$$

$$k_2 = \frac{\dfrac{1}{\tau_d}}{\left(\dfrac{1}{\tau_d} - \dfrac{1}{\tau_{pr}}\right) \cdot \left(\dfrac{1}{\tau_i} - \dfrac{1}{\tau_d}\right)^2} \tag{1.33}$$

$$k_3 = \frac{\dfrac{1}{\tau_{pr}}\dfrac{1}{\tau_d} - \left(\dfrac{1}{\tau_i}\right)^2}{\left(\dfrac{1}{\tau_i} - \dfrac{1}{\tau_{pr}}\right)^2 \cdot \left(\dfrac{1}{\tau_i} - \dfrac{1}{\tau_d}\right)^2} \tag{1.34}$$

$$k_4 = \frac{\left(\dfrac{1}{\tau_{pr}} + \dfrac{1}{\tau_d} - \dfrac{1}{\tau_i}\right) \cdot \left(\dfrac{1}{\tau_i}\right)^2 - \dfrac{1}{\tau_{pr}} \cdot \dfrac{1}{\tau_d} \cdot \dfrac{1}{\tau_i}}{\left(\dfrac{1}{\tau_i} - \dfrac{1}{\tau_{pr}}\right)^2 \cdot \left(\dfrac{1}{\tau_i} - \dfrac{1}{\tau_d}\right)^2} \tag{1.35}$$

Using Equations 1.29 through 1.35, the values of all the shaper-model passive elements are selected.

1.5 OPERATIONAL TRANSCONDUCTANCE AMPLIFIER–BASED SHAPER APPROACH

The main problem in the design of the very large scale integration (VLSI) shaping filters for nuclear spectroscopy is the implementation of long shaping times in the order of microseconds for which high-value resistors (in the megaohms range) and/or capacitors (in the 100-pF range) are demanded. In fact, the practical values in terms of the occupied area that can be integrated are in the range of tens of kiloohms for the resistors and in the picofarads range for the capacitors.

A few examples of monolithic shaping structures providing long shaping times can be found in the literature. In all of the above topologies, different techniques are used in order to obtain relatively large peaking times within a reasonable

occupied area. Specifically, Chase et al. [30] have suggested the use of demultiply-
ing current mirrors to increase the resistor value that is responsible for setting the
filtering time constant of the operational amplifier (op amp)-based low-pass filter.
De Geronimo et al. [31] have also proposed a versatile architecture using current
mirrors to magnify the resistances and that is still based on voltage-mode op amps
as gain blocks. Finally, Bertuccio et al. [32] adopted a current-mode approach to
design a first-order low-pass filter avoiding the use of op amps that result in a very
compact topology.

In this chapter, an alternative design technique is described, and an IC S-G shaping
structure is suggested. The proposed topology, in controversy to the typical shaping
structures, is not based on op amps, which generally demand large-area input transistors and high-bias currents, but rather on operational transconductance amplifiers
(OTAs). An extended study in all the available OTA-based configurations, both in
voltage and current domain, was performed at Refs. [28,29]. All the possible implan-
tations were investigated using all the available advanced filter design techniques.

In this chapter, a Leapfrog OTA-based architecture is described [21]. Although it
is a low-frequency-region-operating bandwidth, it is fully integrated and is characterized by low-power and -noise performance that is compatible with the stringent
requirements of high-resolution nuclear spectroscopy. However, the main advantage
of the specific shaper design is its continuous time-adjustable operating bandwidth,
which renders it suitable for a variety of readout applications.

In particular, respective to Figure 1.6, the two-port passive element network was
designed in order to implement an equivalent fully integrated second-order S-G
shaping filter. This two-port passive network is shown in Figure 1.7.

The specific passive element topology has a respective transfer function (Laplace
representation) to the typical shaper model. Its Laplace representation transfer function for a second-order S-G shaper is given by

$$H(s) = \frac{\left(\dfrac{1}{LC_1}\right)s}{s^3 + s^2(R_S R_L C_1 + L)\dfrac{1}{LC_1} + s\left(\dfrac{C_1}{C_2} + R_L + R_S\right)\dfrac{1}{LC_1} + \dfrac{1}{LC_1 C_2}} \tag{1.36}$$

From the two-port network shown in Figure 1.7, the signal flow graph (SFG) of a
second-order S-G shaping filter is extracted (Figure 1.8).

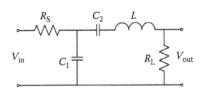

FIGURE 1.7 Equivalent RLC minimum-inductance two-port circuit of a second-order S-G
shaper.

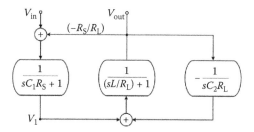

FIGURE 1.8 Signal flow graph of a voltage-mode second-order S-G shaper.

The output signal of the passive network, in relation to the SFG in Figure 1.8, is

$$V_{out} = \frac{1}{\frac{sL}{R_L}+1}\left(V_1 - V_{out}\frac{1}{sC_2R_L}\right) \tag{1.37}$$

Using the extracted SFG and the Leapfrog (functional simulation method) design methodology, a second-order shaper is designed. The main advantage of the Leapfrog method over other filter design methods, which also provide integrated structures, is better sensitivity performance and the capability to optimize the dynamic range by properly intervening during the phase of the original passive synthesis [33,34]. In order to implement the shaping filter, OTAs were selected as the basic building cells. The symbol of the OTA is shown in Figure 1.9.

The OTA is assumed to be an ideal voltage-controlled current source and can be described by

$$I_0 = g_m(V^+ - V^-) \tag{1.38}$$

where I_0 is the output current, and V^+ and V^- denote the noninverting and inverting input voltages of the OTA, respectively. Note that g_m (transconductance gain) is a function of the bias current, I_A. The implementation with OTAs of the S-G shaping filter is greatly advantageous since it provides programmable characteristics. In particular, tunability is achieved by replacing the RC and CR sections in the original passive model with active g_m-C sections, where the g_m can be adjusted with an

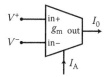

FIGURE 1.9 Symbol of the OTA.

external bias voltage or current. The second-order shaping filter that was designed using the above SFG and the Leapfrog method is shown in Figure 1.10.

The capacitor values and the OTA transconductances of the above S-G shaper are given in Table 1.1. The shaper configuration was designed in order to provide a peaking time that is equal to 1.8 μs, which refers to a bandwidth of 260 kHz in the low-frequency region (f_{c1} = 140 Hz). The passive element values of the respective RLC equivalent two-port network of Figure 1.7 are also given in Table 1.1.

Because of the nonintegrable value of C_2 in the Leapfrog shaper, the specific capacitance was substituted with a grounded OTA-C capacitor simulator. The respective OTA architecture is described in Figure 1.11, and its calculated value is given by Equation 1.39 [33,34]. The values of the transconductances and the capacitor C are $g_{m4} = g_{m5} = g_{m6} = 74.4$ μA/V and $g_{m7} = 950$ nA/V and $C = 12.8$ pF.

$$C_{eq} = C \frac{g_{m4}g_{m5}}{g_{m6}g_{m7}} \qquad (1.39)$$

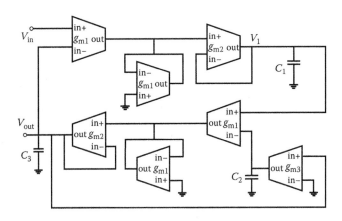

FIGURE 1.10   OTA-based second-order S-G shaper using the Leapfrog technique.

TABLE 1.1

IC Shaper and RLC Equivalent Network Element Values

IC Shaper		Discrete RLC Network	
Active and Passive Elements		**Passive Elements**	
g_{m1}	23.8 μA/V	Rs	100 kΩ
g_{m2}	11.5 μA/V	R_L	100 kΩ
g_{m3}	950 nA/V	C_1	10.33 pF
C_1	12.36 pF	C_2	7.44 nF
C_2	1.00 nF	L	83.7 mH
C_3	13.77 pF		

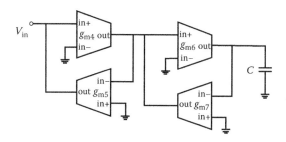

FIGURE 1.11   OTA-based architecture for grounded capacitance simulation.

1.6   ADVANCED CSP–SHAPER ANALOG SIGNAL PROCESSOR IMPLEMENTATION

Using the analysis and methodology presented in Sections 1.2.1 and 1.2.2 on the CSP and shaper design, respectively, an advanced, fully integrated radiation-detection analog processing channel was designed and fabricated [29]. The preamplifier–shaper readout front-end system was designed, simulated, and fabricated in a 0.35-μm CMOS process (3*M*/2*P* 3.3/5 V) by Austria Mikro Systeme for a specific low-energy x-ray silicon strip detector.

The design specifications are given in Table 1.2 [29].

The preamplifier was designed using the architecture that was provided in Figure 1.3. The preamplifier transistor's length and width area are provided in Table 1.3. The transistor dimensions and biases for this configuration were selected according to the design and optimization criteria that were presented in Section 1.2.1 and are provided in Table 1.3. The level-shifting stage was added as the DC bias level of the signal that is provided to the shaper to be controlled and the general dynamic range and 1-dB compression point of the filter that is to be optimized. The feedback capacitance C_f is 550 fF and is placed between the input node and the gate of the source follower stage to avoid introduction on the closed loop of the follower stage complex

TABLE 1.2
Design Specifications

Detector Diode (Si)	
Detector capacitance	C_d = 2–10 pF
Leakage current	I_{leak} = 10 pA
Q collected per event	Q = 312,500 e$^-$
Time needed for the collection of 90% of total Q	300 ns
Preamplifier–Shaper	
Shaper's order	$n = 2$
Shaper peaking time	$\tau_s = n\tau_0 = 1.81$ μs
Temperature	25°C
Power consumption per channel	<8 mW

TABLE 1.3
CSP-Level Shifter MOS Transistor
Dimensions [29]

MOSFET	(W/L) in μm
M1	310/0.9
M2	100/0.7
M3	25/1.5
M4, M5	2.5/5
M6, M7, M9	2.5/2.5
M8	5/2.5
M10	100/0.7
M11	20/1
M12, M13	5/5

poles and to isolate the C_f from the following stage. The bias current I_{bias} was selected to be 10 μA. The power supplies are $V_{DD} = -V_{SS} = 1.65$ V, and the reset device bias voltage is fixed to 150 mV. The level-shifting bias voltage is equal to −1.18 V [29].

The shaper was implemented with the Leapfrog architecture and the OTA-based grounded capacitance simulator replacement structure, depicted in Figures 1.10 and 1.11. An inherent drawback of the Leapfrog methodology and consequently of the particular shaper architecture is that the Leapfrog design method provides a gain value of 1/2 [33,34]. In order to cope with this specific loss, an additional gain stage was used after the preamplifier and before the shaper using an OTA-based structure ($g_{m8} = 153$ μA/V, $g_{m9} = 11.5$ μA/V) that is provided in Figure 1.12.

The specific topology gives the capability to program externally the gain by fixing the bias voltage of the two OTA components. In particular, the DC gain of the above OTA structure is $A = V_{out}/V_{in} = g_{m8}/g_{m9}$. Considering the S-G OTA-based shaping architecture, capacitor simulator, and amplification topology of Figures 1.10, 1.11, and 1.12, respectively, they were implemented using the CMOS OTA configuration, shown in Figure 1.13.

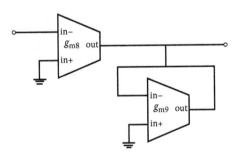

FIGURE 1.12   OTA-based amplification stage.

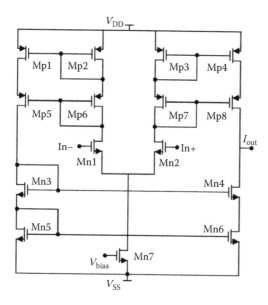

FIGURE 1.13 CMOS OTA.

In the transconductance circuit, a typical CMOS cascode configuration is used, where changing the bias voltage results in approximately equal changes for both the transconductance and the 3-dB frequency. All the OTA circuits' MOS dimensions are given in Table 1.4.

The PMOS devices in the OTA circuits have their bulk biased in the same voltage to their source (designing separate wells), taking advantage the specific n-well CMOS process that is used, in order to avoid the body effect and to maximize the signal swing. The power supplies in all the OTA components are also $V_{DD} = -V_{SS} = 1.65$ V. The bias voltage of each OTA is fixed in -0.85 V in order to achieve the desired transconductance values. The total simplified block diagram of the analog readout ASIC is given in Figure 1.14.

TABLE 1.4

OTA MOSFET Dimensions (μm)

Transistors	g_{m1}	$g_{m2,9}$	$g_{m3,7}$	$g_{m4,5,6}$	g_{m8}
Mp1,Mp4	25/1	25/2.5	25/35	77/1	80/1
Mp2,Mp3	25/1	25/1	25/1	25/1	26/1
Mp5,Mp8	12/1	12/2.5	12/35	36/1	36/1
Mp6,Mp7	12/1	12/1	12/1	12/1	12/1
Mn1,Mn2	3.7/5	3.7/5	3.7/5	3.7/5	3.7/5
Mn3,Mn4	3.7/5	3.7/12.5	2/100	11/5	11/5
Mn5,Mn6	5/5	5/12.5	3/100	15/5	16/5
Mn7	5/5	5/5	5/5	5/5	5/5

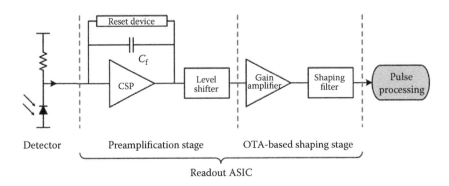

FIGURE 1.14   Block diagram of the IC readout system.

1.7   EXPERIMENTAL RESULTS

The fabricated chip where the prototype x-ray readout ASIC was implanted is shown
in Figure 1.15. The measured x-ray IC front-end system output signal is shown in
Figure 1.16. The system provides a DC gain that is equal to 120 dB. However, as it
will be presented in Figure 1.15, this value can be externally modified by suitably
fixing the bias voltages and consequently changing the g_m values of the OTA-based
amplification topology of Figure 1.12. Regarding the radiation-detection application
specifications, the output pulse has a peaking time value of 1.81 μs that shows no
undershoot or pileup.

The power consumption is 1 mW, far lower than the maximum allowed specified
value of 8 mW, rendering the system as low power [24,35].

The readout ASIC noise performance was also analytically studied. The system
ENC is 382 e– for a detector of 2 pF, and the noise performance increases with a

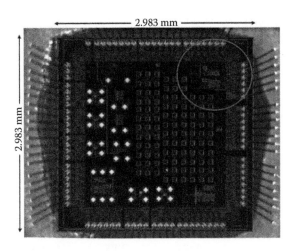

FIGURE 1.15   Microphotograph of the full-fabricated microchip. (The readout system is
circled.)

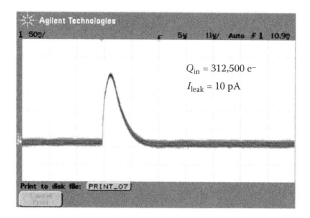

FIGURE 1.16 Readout system output pulse.

slope of 21 e⁻/pF. The ENC dependence on the detector capacitance variations is shown in Figure 1.17.

Considering the front-end system energy resolution–linearity, it is presented in Figure 1.18. The CMOS readout analog processor achieves an input charge gain–voltage output conversion of 3.31 mV/fC and a linearity of 0.69%.

In terms of the total occupied area, the specific IC system is again advantageous since it consumes only 0.2017 mm^2 of a total 2.983 × 2.983-mm microchip [36,37].

Finally, the system flexibility to be used in a wide range of readout applications, taking advantage of the specific shaping stage topology, was examined. The system gain is programmable from 118 to 137 dB by changing the OTA's bias voltage of the gain stage. This gain programmability is demonstrated in Figure 1.19. Furthermore, the peaking time can similarly be externally modified. This is represented in Figure 1.20, where output signals with peaking times from 1 to 3 μs are provided by suitably

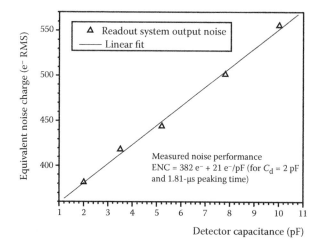

FIGURE 1.17 Readout ASIC ENC dependence on the detector capacitance.

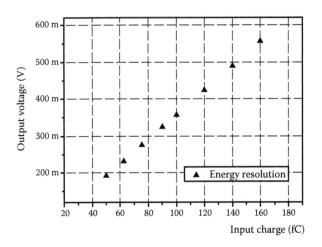

FIGURE 1.18 Readout ASIC measured energy response.

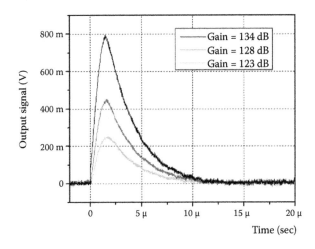

FIGURE 1.19 Gain programmability.

fixing the bias voltages of the OTA configurations, which operate in the total shaper architecture as integrators and the OTAs in the capacitance simulator unit. As the peaking time increases, the shaper upper frequency becomes higher in its bandpass response. The particular OTA-based S-G shaper provides continuously variable peaking time in the range of 950 ns–3.1 μs. A variation of approximately 30% in the signal amplitude can be detected when moving from the shortest to longest available peaking time. The output signal undershoot can also be externally adjusted. In Figure 1.21, the different output signal undershoots are shown. The undershoots can vary from 0% to 18% of the positive signal amplitude. This results to a respective increment of the low 3-dB frequency providing a narrower shaper bandwidth. Consequently, the output noise can also be regulated in relation to each application.

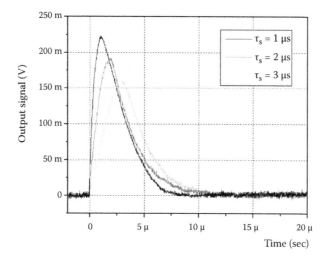

FIGURE 1.20 Adjustable peaking time capability–output signal for different peaking times.

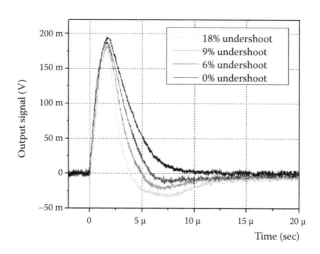

FIGURE 1.21 Variable output signal undershoot capability—different output signal undershoot.

Regarding the system and particularly the S-G shaping filter stability, having only one feedback between its input and output, no stability problems were observed. In addition, no sensitivity problems occurred in relation to the g_m terms since their value could be fixed externally using the respective bias voltages.

1.8 CURRENT CHALLENGES AND LIMITATIONS

Radiation-detection IC design imposes new challenges and limitations as the technology nodes are shrinking.

In terms of the technology used, the CMOS has scaled down to 28 and 20 nm allowing extremely dense integration and cost minimization for high-volume production. However, in terms of noise optimization, performance, and cost, the process selection should always be defined by the related application. While the noise becomes lower in submicron technologies, extra noise sources should be taken into account like the gate-induced noise and the channel-reflected noise. In terms of the circuit design, advanced noise optimization methodologies are needed not only in the circuit architecture level but also on the mask design (physical layout) level, as these noise contributors need to be optimized. While in previous technology nodes, these were almost negligible, now, in 28 nm and below, these may contribute up to 50% of the noise depending always on the application and mostly on the operating circuit/system frequency bandwidth.

In addition, the related transistor noise models, always in relation to the used technology Fab, are not adequate to provide an accurate simulation all the time, and the related performance cannot be effectively optimized. Advanced models are needed as the designer to be able to access the trade-offs and to optimize the radiation-detection IC processor.

Furthermore, in terms of radiation hardness, extreme limitations exist. Models for pre- and postradiation conditions are needed as the performance to be accessed in real operation mode. No technology foundry provides this kind of simulation support, and, as a result, the related performance degradation after irradiation cannot be estimated effectively in all the presilicon design stages and verification. Advanced modeling techniques are required, and the respective device layouts for radiation-hard operation are demanded as the radiation IC to be able to cover the operation specifications under a radiated environment. This is a gap in the design flow that renders the radiation IC design extremely risky at all times.

REFERENCES

1. Beuville, E., Borer, K., Chesi, E., Heijne, E.H.M., Jarron, P., Lisowski, B. and Singh, S. 1990. Amplex, a low-noise, low-power analog CMOS signal processor for multi-element silicon particle detectors. *Nuclear Instruments and Methods in Physics Research A* 288(1), pp. 157–167.
2. Wurtz, L.T. and Wheless, Jr., W.P. 1993. Design of a high performance low noise charge preamplifier. *IEEE Transactions on Circuits and Systems I* 40(8), pp. 541–545.
3. Tedja, S., Van der Spiegel, J. and Williams, H.H. 1995. A CMOS low-noise and low-power charge sampling integrated circuit for capacitive detector/sensor interfaces. *IEEE Journal of Solid State Circuits* 30(2), pp. 110–119.
4. Radeka, V., Rehak, P., Rescia, S., Gatti, E., Longoni, A., Sampietro, M., Holl, P., Struder, L. and Kemmer, J. 1988. Design of a charge sensitive preamplifier on high resistivity silicon. *IEEE Transactions on Nuclear Science* 35(1), pp. 155–159.
5. Radeka, V., Rehak, P., Rescia, S., Gatti, E., Longoni, A., Sampietro, M., Bertuccio, G., Holl, P., Struder, L. and Kemmer, J. 1989. Implanted silicon JFET on completely depleted high-resistivity devices. *IEEE Electron Devices Letters* 10(2), pp. 91–94.
6. Lund, J.C., Olschner, F., Bennett, P. and Rehn, L. 1995. Epitaxial n-channel JFETs integrated on high resistivity silicon for x-ray detectors. *IEEE Transactions on Nuclear Science* 42(4), pp. 820–823.

7. Lechner, P., Eckbauer, S., Hartmann, R., Krisch, S., Hauff, D., Richter, R., Soltau et al. 1996. Silicon drift detectors for high resolution room temperature x-ray spectroscopy. *Nuclear Instruments and Methods A* 377, pp. 346–351.

8. Ratti, L., Manghisoni, M., Re, V. and Speziali, V. 2001. Integrated front-end electronics in a detector compatible process: Source-follower and charge-sensitive preamplifier configurations. In: James, R.B. ed., Hard x-ray and gamma-ray detector physics. *III Proceedings of SPIE* 4507, pp. 141–151.

9. Sansen, W. and Chang, Z.Y. 1990. Limits of low noise performance of detector readout front ends in CMOS technology. *IEEE Transactions on Circuits and Systems* 37(11), pp. 1375–1382.

10. Chang, Z.Y. and Sansen, W. 1991. Effect of 1/f noise on the resolution of CMOS analog readout systems for microstrip and pixel detectors. *Nuclear Instruments and Methods* 305, pp. 553–560.

11. Steyaert, M., Chang, Z.Y. and Sansen, W. 1991. Low-noise monolithic amplifier design: Bipolar versus CMOS. *Analog Integrated Circuits and Signal Processing* 1(1), pp. 9–19.

12. Motchenbacher, C.D. and Connelly, J.A. 1998. *Low-Noise Electronic System Design.* Wiley Interscience Publications, New York.

13. Xie, D., Cheng, M. and Forbes, L. 2000. SPICE models for flicker noise in n-MOSFETs from subthreshold to strong inversion. *IEEE Transactions on Computer-Aided Design* 19(11), pp. 1293–1303.

14. Noulis, T., Siskos, S. and Sarrabayrouse, G. 2007. Comparison between BSIM4.X and HSPICE flicker noise models in NMOS and PMOS transistors in all operating regions. *Microelectronics Reliability* 47, pp. 1222–1227.

15. Jakobson, C.G., Bloom, I. and Nemirovsky, Y. 1998. 1/f noise in CMOS transistors for analog applications from subthreshold to saturation. *Solid-State Electronics* 42(10), pp. 1807–1817.

16. Nemirovsky, Y., Brouk, I. and Jakobson, C.G. 2001. 1/f noise in CMOS transistors for analog applications. *IEEE Transactions on Electron Devices* 48(5), pp. 921–927.

17. Liu, W. 2001. *MOSFET Models for SPICE Simulation Including BSIM3V3 and BSIM4.* Wiley–Interscience, New York.

18. Noulis, T., Siskos, S. and Sarrabayrouse, G. 2008. Noise optimized charge sensitive CMOS amplifier for capacitive radiation detectors. *IET Circuits Devices and Systems* 2(3), pp. 324–334, June.

19. Chang, Z.Y. and Sansen, W. 1991. *Low Noise Wide Band Amplifiers in Bipolar and CMOS Technologies.* Kluwer, Norwell MA.

20. Noulis, T., Deradonis, C., Siskos, S. and Sarrabayrouse, G. 2006. Programmable OTA based CMOS shaping amplifier for x-ray spectroscopy. *Proceedings of 2nd IEEE PRIME,* Otranto (Lecce), Italy, pp. 173–176.

21. Noulis, T., Siskos, S., Sarrabayrouse, G. and Bary, L. 2008. Advanced low noise x-ray readout ASIC for radiation sensor interfaces. *IEEE Transactions on Circuits and Systems,* Part 1, 55(7), pp. 1854–1862, August.

22. Vandenbussche, J., Leyn, F., Van der Plas, G., Gielen, G. and Sansen, W. 1998. A fully integrated low-power CMOS particle detector front-end for space applications. *IEEE Transactions on Nuclear Science* 45(4), pp. 2272–2278.

23. Guazzoni, C., Samprieto, M. and Fazzi, A. 2000. Detector embedded device for continuous reset of charge amplifiers: choice between bipolar and MOS transistor. *Nuclear Instruments and Methods in Physics Research A* 443, pp. 447–450.

24. Pedrali-Noy, M., Gruber, G., Krieger, B., Mandelli, E., Meddeler, G., Moses, W. and Rosso, V. 2001. PETRIC-A position emission tomography readout integrated circuit. *IEEE Transactions on Nuclear Science* 48(3), pp. 479–484.

25. Konrad, M. 1968. Detector pulse-shaping for high resolution spectroscopy. *IEEE Transactions on Nuclear Science* 15(1), pp. 268–282.
26. Ohkawa, S., Yoshizawa, M. and Husimi, K. 1976. Direct synthesis of the Gaussian filter for nuclear pulse amplifiers. *Nuclear Instruments and Methods* 138(1), pp. 85–92.
27. Goulding, F.S. and Landis, D.A. 1982. Signal processing for semiconductor detectors. *IEEE Transactions on Nuclear Science* 29(3), pp. 1125–1141.
28. Noulis, T., Deradonis, C., Siskos, S. and Sarrabayrouse, G. 2007. Detailed study of particle detectors OTA based CMOS semi-Gaussian shapers. *Nuclear Instruments and Methods in Physics Research A* 583, pp. 469–478.
29. Noulis, T., Deradonis, C. and Siskos, S. 2007. Advanced readout system IC current mode semi-Gaussian shapers using CCIIs and OTAs. *VLSI Design Journal*, 2007, article ID 71684, 12 pp.
30. Chase, R.L., Hrisoho, A. and Richer, J.P. 1998. Eight-channel CMOS preamplifier and shaper with adjustable peaking time and automatic pole-zero cancellation. *Nuclear Instruments and Methods A* 409, pp. 328–331.
31. De Geronimo, G., O'Connor, P. and Grosholz, J. 2000. A generation of CMOS readout ASIC's for CZT detectors. *IEEE Transactions on Nuclear Science* 47, pp. 1857–1867.
32. Bertuccio, G., Gallina, P. and Sampietro, M. 1999. "R-lens filter": An (RC)^n current-mode low pass filter. *Electronics Letters* 35, pp. 1209–1210.
33. Deliyannis, T., Sun, Y. and Fidler, K. 1999. *Continuous-Time Active Filter Design.* CRC Press LLC, Florida.
34. Sedra, A.S. and Brackett, P.O. 1979. *Filter Theory and Design: Active and Passive.* Pitman, London.
35. Krieger, B., Ewell, K., Ludewigth, B.A., Maier, M.R., Markovic, D., Milgrome, O. and Wang, Y.J. 2001. An 8 × 8 pixel IC for x-ray spectroscopy. *IEEE Transactions on Nuclear Science* 48(3), pp. 493–498.
36. Limousin, O., Gevin, O. Lugiez, F., Chipaux, R., Delagnes, E., Dirks, B. and Horeau, B. 2005. IDeF-X ASIC for Cd(Zn)Te spectro-imaging systems. *IEEE Transactions on Nuclear Science* 52(5), pp. 1595–1602.
37. Shani, G. 1996. *Electronics for Radiation Measurements*, vol. 1. CRC Press LLC, Florida, pp. 182–183.

2 Analog Electronics for HVCMOS Sensors

Ivan Peric

CONTENTS

2.1 CMOS Active Pixel Sensors 27
2.2 Triple-Well MAPS 29
2.3 INMAPS 29
2.4 HVCMOS Detectors 30
2.5 HVCMOS Development 33
2.6 Design Details 34
    2.6.1 Bias Resistance 35
    2.6.2 Amplifier 38
    2.6.3 Resistive Feedback as Continuous Reset 39
    2.6.4 Passive Shaper and Threshold Control 39
    2.6.5 Discriminator and Latch 40
2.7 New Developments 41
2.8 Mu3e 41
2.9 MuPix 42
References 46

2.1 CMOS ACTIVE PIXEL SENSORS

The first CMOS-based sensor chips were developed in the 1960s, but the world of solid-state imaging was revolutionized by another invention. In 1970, charge-coupled devices (CCDs) were invented at the Bell laboratory. Very soon, CCDs took over all the competing technologies. For approximately 20 years, from the invention of CCDs to the late 1980s, CMOS sensors were confined to very specialized applications, namely, infrared focal-plane detectors [1]. In the past decade, however, CMOS sensors gained popularity again due to the extremely fast development of CMOS technologies, which gives the possibility to implement large-area CMOS chips with many millions of components and to implement very small transistors and therefore small pixels. Further, the pixel matrix and readout circuits can be implemented on the same chip, which allows high parallelization of the readout and improves readout speed.

In the case of a typical CMOS active pixel sensor (APS) for visible-light detection in digital cameras, every pixel contains a reverse-biased photosensitive diode (n-well) and a few transistors, which are used to amplify, clear, and read out the signal. Due to these additional devices, the surface of a pixel is not 100% sensitive to light, and a portion of the incident light is not detected. This signal loss leads to a slight decrease of the light sensitivity of a camera.

The use of CMOS sensors in particle physics was proposed in 1999. The main difference with respect to visible-light applications is that the sensor has to be 100% efficient. Due to local generation of the signals in the case of a CMOS photosensor (the signals are generated in the depleted area of the photosensitive diode), many particles will not be detected.

In order to solve this problem, several groups have been working on the development of monolithic active pixel sensors (MAPSs). These sensors are based on the following principle: in modern CMOS processes, the n- and p-well housing the transistors are usually fabricated on the top of the thin (5–25-μm), p-doped, epitaxially grown *epilayer*. The typical resistivity of this layer is ~10 Ωcm. A small n-well in the p-epilayer is used for charge collection. Because of the lower doping of the epilayer with respect to the p-wells above it and the p-substrate below, the epilayer acts as a shallow potential well for electrons. Electrons created by radiation diffuse in the epilayer and spread over a relatively large area. Some of them get close enough to an n-well/p-epi diode where they experience the electric field and get collected.

In this way, it is possible to make compact and cheap detectors with a 100% fill factor. Due to the possibility of implementing small pixels and linear charge sharing, spatial resolution is excellent.

Figure 2.1 shows the cross section through a standard monolithic CMOS sensor.

Unfortunately, MAPSs in their original form have a few limitations. First, not every semiconductor technology is suitable for MAPSs. Second, the charge collection by diffusion is relatively slow. Because of the slow charge collection, the radiation tolerance is relatively poor. The charge carriers get easily trapped or recombined in the epilayer that is damaged by radiation. Further, only n-channel metal oxide semiconductor (NMOS) transistors can be used inside pixels; a p-channel metal oxide semiconductor (PMOS) transistor would lead to signal loss since its positively biased n-well would collect a large part of the signal charge. This is illustrated in Figure 2.2.

The best property of MAPSs is their excellent spatial resolution (a single-point resolution of 1.5 μm for 20-μm and ~1 μm for 10-μm pixel pitch has been demonstrated

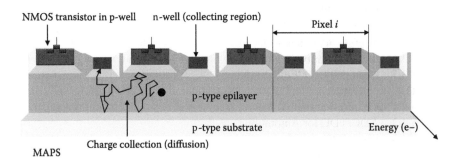

FIGURE 2.1 Monolithic active pixel sensor. Charge collection relies on diffusion.

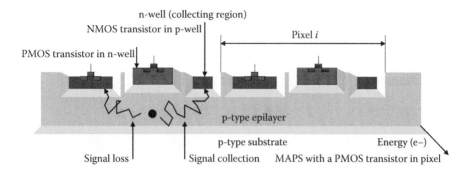

FIGURE 2.2   Introducing of a PMOS transistor in a MAPS pixel would lead to a charge loss.

[2,3]), which makes them a good technology choice for precise vertex detectors, particularly for flavor tagging.

In addition to their applications as visible-light sensors and as particle trackers, the MAPSs are (or can be) used in x-ray photoelectron emission microscopy or electron microscopy and as a photosensitive part of indirect x-ray detectors.

There have been a few attempts to improve the MAPS concept and allow the implementation of CMOS circuits inside pixels.

2.2   TRIPLE-WELL MAPS

Triple-well MAPSs rely as the standard MAPSs on the charge generation in the undepleted epilayer (or silicon substrate) and the charge collection by diffusion. However, a much larger n-well is used as the collection region. This n-well is almost as large as the pixel itself. A technology is chosen that allows the implementation of p-wells housing NMOS transistors inside the collecting n-well. PMOS transistors are placed in a separate and much smaller (secondary) n-well. The use of a large n-well as a collecting region improves signal collection speed and thus the radiation tolerance. However, the secondary n-well collects a part of the charge, which leads to a signal loss. Because of this, the number of PMOS transistors must be minimized. These detectors have a large but not perfect fill factor. A schematic cross section through a triple-well MAPS is shown in Figure 2.3.

2.3   INMAPS

Isolated n-well monolithic active pixel sensors (INMAPS) detectors can implemented in a CMOS process with an epilayer that is modified by adding a deep p-implant. INMAPS detectors have a similar structure as the standard MAPSs; they also rely on diffusion as a signal collection mechanism. Different from a standard MAPS, an INMAPS detector can use both PMOS and NMOS transistors inside pixels. PMOS transistors are shielded from the epilayer by the deep p-well, which prevents signal loss. A schematic cross section through an INMAPS is shown in Figure 2.4.

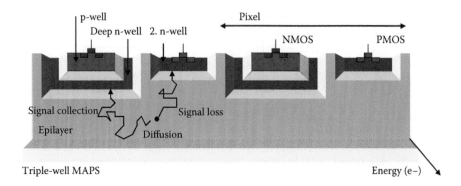

FIGURE 2.3   Schematic cross section through a triple-well MAPS.

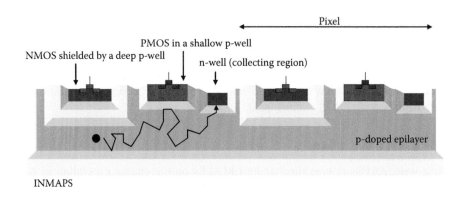

FIGURE 2.4   Schematic cross section through an INMAPS.

2.4 HVCMOS DETECTORS

As mentioned in Section 2.1, MAPS sensors in their original form have a few limitations; particularly, the charge collection by diffusion is relatively slow, and the radiation tolerance is poor. Further, only NMOS transistors can be used inside pixels, which limits the in-pixel electronics. A few improved MAPS concepts have been introduced by several groups and collaborations to overcome these drawbacks. Some of the improved concepts are based on technology adjustments, such as INMAPS. Another concept—triple-well MAPS—is based on the use of a larger collecting region—a deep n-well. The common factor for all the MAPS variants is that they are based on signal generation in an undepleted silicon substrate or epilayer.

High-voltage monolithic detectors can be implemented in HVCMOS technologies. These technologies are often used by industry. Their main applications are power management circuits for mobile phones; automotive bus transceivers; printer head electronics; and liquid crystal display (LCD)-, motor-, and data line drivers for high-speed internet telephony (*voice over IP*). HVCMOS technologies allow both the implementation of special transistors that are capable of generating high-voltage signals (up to 120 V) and the integration of standard low-voltage transistors that are commonly used to control high-voltage devices.

The key element of an HVCMOS sensor is the *floating logic* structure. A group of PMOS and NMOS transistors can be electrically isolated from a lightly doped p-type substrate by a high-voltage deep n-well. (PMOS transistors are placed in the shallow n-wells that are ohmically connected with the deep n-well; NMOS transistors are in the p-wells, which are inside the deep n-well. Such a structure is originally used as an interface between the low- and the high-voltage circuits.) Depending on the technology used, the deep n-well/p-substrate junction can sustain a reverse bias of up to 120 V. Depending on substrate resistivity, a depleted zone in the range of 15–100 μm can be induced around the n-well. The signals generated in the depleted zone are collected by drift.

The HVCMOS detector is based on two main ideas.

The first idea is to use the deep n-well as the signal collection region and the depleted p-substrate/n-well junction as the sensor. The second idea is to implement the entire pixel electronics with both PMOS and NMOS transistors inside the deep n-well. Since CMOS electronics can be placed inside the n-well, any type of complex signal processing electronics can be implemented inside the cathode of the sensor diode (n-well).

Further, if the deep n-wells are arranged as a matrix, their depleted zones will partially overlap. Wherever the charge signal is generated below or between the n-wells, it will be collected to the nearest n-well. The pixel detector therefore has a nearly 100% fill factor. The detector structure is shown in Figure 2.5. Since the signal electrons experience a strong electric field, they will be collected very fast, and a high radiation tolerance can be expected.

The main complication in the concept is that the deep n-well has two roles. First, the deep n-well is the signal-collecting region. Second, the deep n-well is the substrate for PMOS transistors that are placed in it. Usually, an n-well utilized as a PMOS transistor substrate is connected to the positive supply voltage, which makes its potential stable. Here, we cannot bias the n-well in this way because the signal charge would be absorbed by the positive supply. The n-well has to be *floating* for fast-charge signals, i.e., it can be biased only by using a high resistance or a reset switch. In this way, we obtain a very unusual structure. Pixel electronics (particularly its PMOS part) have to sense and amplify the signal-induced voltage drop in its own substrate. Still, that concept works well. However, attention has to be paid on capacitive coupling from signals that are generated by in-pixel electronics to the n-well.

Let us compare the high-voltage sensor with, at first sight, the similar triple-well MAPS (Figure 2.6). A triple-well MAPS also uses a deep n-well as a collection region. This is actually the only common thing between the concepts. Triple-well MAPSs do not use the collecting n-well as the substrate for PMOS transistors, but another secondary well is used for this purpose, which leads to a certain charge loss. Further, triple-well MAPSs are implemented in low-voltage technologies. The

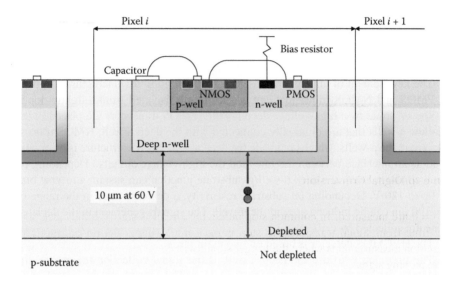

FIGURE 2.5 HVCMOS detector.

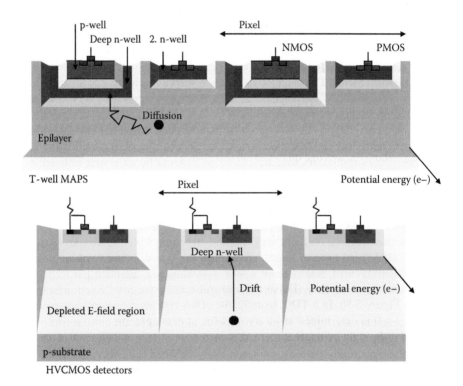

FIGURE 2.6 Comparison between HVCMOS detector and triple-well MAPS.

depleted zone of the n-well/p-substrate junction is in this case too thin to provide a significant signal. These detectors rely on the signals that are generated below the depleted zone and on signal collection by diffusion.

To summarize, an HVCMOS detector in high-voltage technology has the following characteristics:

- It is implemented in a commercial technology without any adjustment.
- Signal generation occurs in the depleted semiconductor: signal collection by drift.
- Both p- and n-channel transistors can be used inside the pixel.
- The detector virtually has a 100% fill factor.

2.5   HVCMOS DEVELOPMENT

It should be mentioned that the development of monolithic detectors in high-voltage technology would not start without the support of the foundation Baden Württemberg Stiftung *"Landes Baden Württemberg Stiftung."* Several test detectors have been designed to prove the principle. The test detectors can by classified into the following two types:

1. *Rolling-shutter detectors*: The pixel electronics are as simple as possible and allows solely selecting pixel rows, analog rolling-shutter readout, and the use of a source follower as an amplifier. Only PMOS transistors are used inside pixels, which save space and allow the design of relatively small collecting n-well electrodes with low detector capacitance (C_{det} ~ 10 fF). The advantages of such rolling-shutter detectors are their small pixel size/capacitance and the absence of static current consumption. As an example of a rolling-shutter detector, we mention HVPixelM. It is implemented in the 0.35-μm HVCMOS technology of the company ams AG; it contains a 128×128-pixel matrix with 21×21-μm^2 pixels. The matrix readout time is 50 μs. A typical signal is 2200 e (cluster). Noise of about 20e has been measured. The layout of four pixels (HVPixelM detector) is shown in Figure 2.7. The photomicrograph of the HVPixelM is shown in Figure 2.8.

2. *Detectors with smart pixels*: The pixel electronics contain a charge-sensitive amplifier with continuous reset, a discriminator, and a threshold-tune digital-to-analog converter (DAC) and allow a fast trigger-based *binary* readout. Both PMOS and NMOS transistors are used inside pixels. The advantages of such detectors are in-pixel signal processing (hit detection), the possibility of signal rise time measurements, and automated leakage current compensation. The pixel size and capacitance are larger (C_{det} ~ 100 fF – 200 fF) despite the fact that the noise values are relatively low due to the absence of pulsed reset and in-pixel noise filtering. (Typically, the equivalent noise charge ~ 60 e for 125-ns shaping time.) The layout of a smart pixel (HVPixel chip) is shown in Figure 2.9. The photomicrograph of the chip is shown in Figure 2.10.

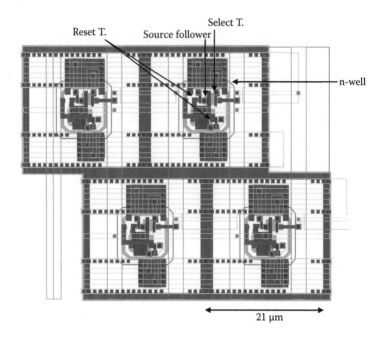

FIGURE 2.7 Layout of four small pixels with rolling-shutter readout (HVPixelM chip). T, transistor.

2.6 DESIGN DETAILS

Nearly all smart pixels implemented in HVCMOS technology are based on the circuit that is implemented on the HVPixel chip (Figure 2.10). Section 2.6 explains this circuit in detail.

Figure 2.11 shows the schematic cross section of a pixel and the block diagram of the pixel electronics as implemented on the test chip. The pixel electronics are completely implemented inside the deep n-well. The p-substrate is biased with a high negative voltage with respect to the n-well. In this way, a large depleted area is generated. The pixel electronics comprises a charge-sensitive amplifier, which is capacitively coupled to a sensor, a continuous reset, a band-pass filter (CR-RC filter), a discriminator, a 4-bit threshold-tune DAC, and a digital latch that stores the hit flag. The latch can be read out using a digital differential bus.

As already mentioned in Figure 2.11, the large deep n-well plays two roles. First, it is the substrate for the PMOS transistors and p-wells. Second, the deep n-well is the cathode of the sensor diode. It is a common practice in the CMOS chip design to bias an n-well that contains PMOS transistors by shorting it with the positive supply. In this way, it is assured that the source and drain diodes of the PMOS transistors are reversely biased. A low-ohmic bias is important to prevent the dangerous latch-up effect—the triggering of the parasitic thyristor, which leads to a high current between the positive supply and the ground. In the case of the pixel structure in Figure 2.11, the shorting of the n-well with the positive supply would lead to a signal loss. The electrons generated

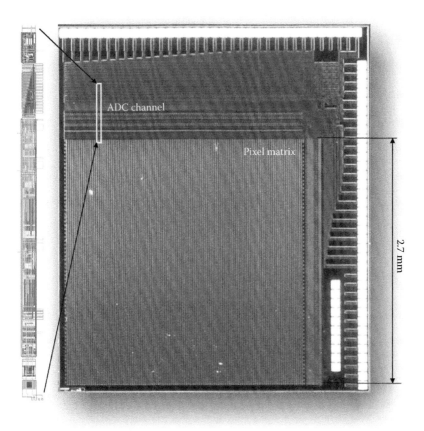

FIGURE 2.8   Photomicrograph of HVPixelM chip and layout of an ADC channel.

by a particle hit would simply flow into the positive supply line before the amplifier can react. To avoid such signal loss, the n-well is connected to the positive supply using a high (typically 1-GΩ) resistance. The charge-sensitive amplifier will suck in and amplify the charge that is generated by a particle. After that, the slow action of the bias resistance restores the initial bias (DC) voltages. The latch-up can be prevented by careful biasing of the p-wells inside the deep n-well and by the use of guard rings.

An additional design problem is the crosstalk. Every PMOS P+ diffusion is capacitively coupled to the deep n-well and therefore to the sensor cathode. Since we are dealing with relatively weak input signals, even the crosstalk generated by a moderate voltage signal on a minimum-size P+ diffusion can overtop the input signal. Great attention has to be paid; we will discuss the possible solutions of the problem in Sections 2.6.4 and 2.6.5.

2.6.1   Bias Resistance

The bias resistance (see Figure 2.11) is implemented using a PMOS transistor, which operates in a linear region (transistor Mb in Figure 2.12). A polysilicon

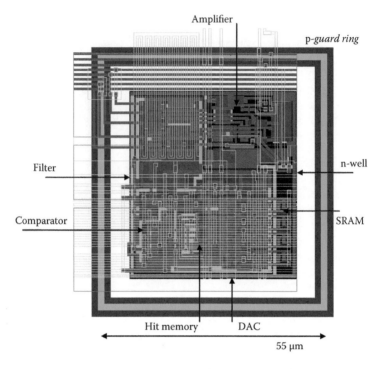

FIGURE 2.9   Layout of a smart pixel (HVPixel chip). SRAM, static random access memory.

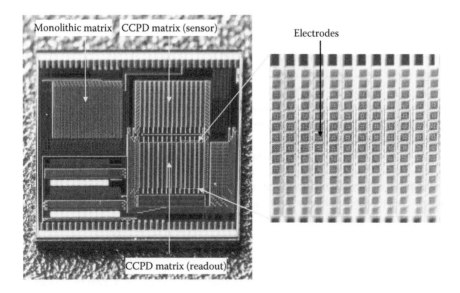

FIGURE 2.10   Photomicrograph of HVPixel chip. CCPD, capacitively coupled pixel detector.

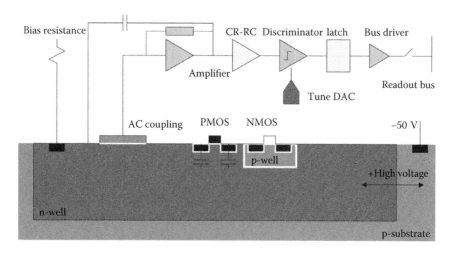

FIGURE 2.11   Block scheme of the pixel.

implementation would require an unreasonably large space and would lead to a large parasitic capacitance. The gate voltage of the transistor Mb can be generated by a diode-connected PMOS transistor that is placed outside the pixel, as shown in Figure 2.12. The bias current can be varied using a bias DAC that is placed on the chip periphery. The bias circuit works even when the DAC current is set to zero. The P+/n-well diodes that have anodes connected to the positive supply (V_{dd}) then act as bias resistors and define the n-well potential. One such diode is depicted in Figure 2.12. The diode-based bias is used in some MAPS implementations with continuous reset and a capacitively- (or AC) coupled sensor. The P+/n-well diodes conduct small-leakage currents. They work at the onset of forward region having high dynamic resistance. They introduce, therefore, very little noise.

In the case of a nonzero bias current, the equivalent resistance of the bias circuit in Figure 2.12 is the parallel connection of the P+/n-well diode dynamic resistances and the resistance of transistor Mb.

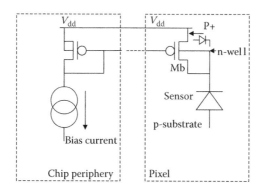

FIGURE 2.12   Bias resistance.

2.6.2 Amplifier

The transistor schematic of the charge-sensitive amplifier is shown in Figure 2.13. The amplifier is AC-coupled to the sensor. AC coupling allows the use of the amplifier types that have input-node DC voltage that is different from the n-well bias potential. In our case, the amplifier has a folded-cascode single-ended topology with a PMOS transistor as an input device. The amplifier is biased with a relatively small current (9 µA).

As already mentioned in Section 2.6, every P+ diffusion (denoted in Figure 2.13 by the letters a, b, c, and d) is capacitively coupled to the n-well and can be a source of unwanted crosstalk. P+ diffusions a and c are shorted to the positive supply and represent solely an additional detector capacitance. The same holds for P+ diffusion b, which is kept by the action of cascode transistor M_c at a nearly constant potential. The P+ diffusion d is the output of the amplifier, and the signals on this node are capacitively coupled to the n-well (sensor). The capacitance between diffusion d and the n-well (c_{fb}) acts as the parasitic capacitive feedback for the charge-sensitive amplifier. The charge-sensitive amplifier is designed without any other feedback capacitor; it relies exclusively on c_{fb} capacitance. The charge gain of the amplifier is $1/c_{fb}$. Since the load transistor (M_l) can be made narrow (there is no need for a large transconductance), diffusion capacitance c_{fb} can be kept small (0.9 fF). In this way, a high charge gain can be achieved.

It is worth mentioning that the parasitic feedback connects the output of the charge-sensitive amplifier with the input prior to the coupling capacitor C_c, see Figure 2.13. The coupling capacitor is therefore included in the feedback loop. Usually, the feedback capacitor of a charge amplifier is placed between its output and the gate of the input transistor. The coupling capacitor is then placed outside the feedback loop. Including the coupling capacitor in the feedback loop has the advantage that there is no signal loss due to charge division between the coupling and sensor capacitance (provided that the gate capacitance of the input transistor is much smaller than the sensor capacitance). The coupling capacitance does not need to be larger than the sensor capacitance.

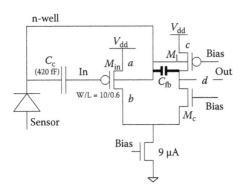

FIGURE 2.13   Charge-sensitive amplifier.

The capacitance of the input transistor is estimated to be approximately 20 fF. This is by one order of magnitude lower than the sensor capacitance (approximately 220 fF). Simulations have shown that increasing the input transistor width does not lead to a better signal/noise ratio if the bias current is limited to a small value ~10 µA. (Input transistor works in weak inversion where the transconductance does not depend on the transistor size.)

2.6.3   RESISTIVE FEEDBACK AS CONTINUOUS RESET

The scheme of the continuous-reset circuit is shown in Figure 2.14. The circuit stabilizes the amplifier and discharges the feedback capacitance after signal integration. For large positive-output signals, the current of the feedback transistor M_{fb} saturates, and the feedback capacitance is discharged with a constant current until the output signal drops below a few thermal voltages ($v_T \sim 26$ mV). A typical output waveform is sketched in Figure 2.14. The bias circuit (shown in Figure 2.14) is used to generate the gate–source voltage for the feedback transistor, which is independent on the DC potential at the input of the amplifier. The bias circuit in Figure 2.14 allows us to change the amplifier bias without affecting the feedback resistance.

2.6.4   PASSIVE SHAPER AND THRESHOLD CONTROL

Figure 2.15 shows the CR-RC shaper, the threshold tune circuit, and the discriminator. The shaper and the discriminator are implemented using only NMOS transistors in order to avoid the capacitive crosstalk of signals to the n-well sensor electrode. The CR stage (devices M_z and C_z) has the purpose to prevent the low-frequency noise passing the shaper. The bias settings of the transistors can be varied using the in-chip DACs; the typical time constant of the CR stage (derived from the small-signal parameters) is 1.3 µs. The RC stage is implemented as a source follower with

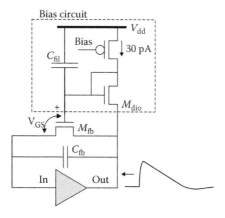

FIGURE 2.14   Continuous-reset circuit.

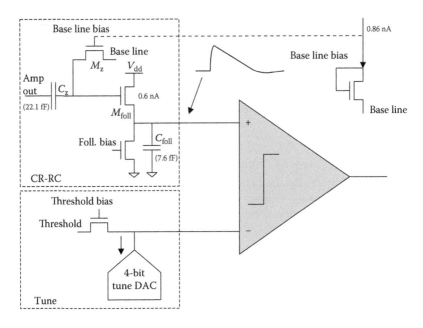

FIGURE 2.15   Shaper, threshold tune circuit, and discriminator.

capacitive load (devices M_{foll} and C_{foll}); its task is to limit the bandwidth of the system and decrease the noise component that is generated by the input transistor of the amplifier. The typical RC stage time constant is 500 ns. Such a long time constant has been chosen to improve the signal/noise ratio. We must point out that using the long shaping time prevents the accurate measurement of the charge collection speed. In order to perform this measurement, the amplifier and shaper have to be optimized for high-speed operation.

A 4-bit tune DAC is implemented as a matrix of 15 identical NMOS current sources. Both transistors M_z and M_{foll} saturate after receiving a large signal, and the voltage at the output of the shaper returns to the base line with constant speed. The pulse width at the output of the discriminator is linearly proportional to the signal amplitude (supposing that the signal amplitude is higher than a few thermal voltages). This gives the possibility to determine the signal amplitude by measuring the discriminator pulse width.

2.6.5 DISCRIMINATOR AND LATCH

The discriminator is implemented as an NMOS-based fully differential amplifier with NMOS diodes as the load (Figure 2.16). The circuit is followed by a differential current-mode latch, which stores the hit flag. Like the discriminator, the latch is realized exclusively with NMOS transistors, as shown in Figure 2.16.

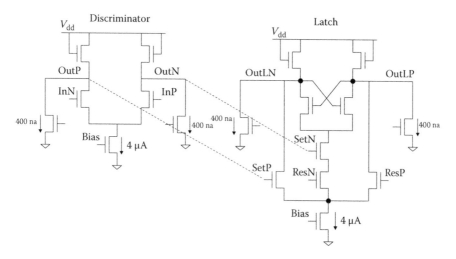

FIGURE 2.16 Discriminator and current-mode latch.

2.7 NEW DEVELOPMENTS

Since 2010, HVCMOS sensors have had a few applications: HVCMOS sensors will be used in the Mu3e experiment at Paul Scherrer Institute (PSI), Switzerland, and are considered as an option for a toroidal LHC apparatus (ATLAS) strip and pixel layers (Large Hardon Collider [LHC]/The European Organization for Nuclear Research [CERN]) and Compact Linear Collider (CLIC) (CERN). The Mu3e detector will be a 50-μm thin detector with an area of 2 m². For CLIC, a capacitively coupled hybrid pixel detector is proposed where the capacitive signal transmission is used for the readout of sensor signals. For ATLAS, there are several detector options, including hybrid and monolithic sensors; the total area of the sensor will be nearly 100 m².

2.8 Mu3e

The aim of the Mu3e experiment at PSI is the search for the particle decay: $\mu^+ \rightarrow e^+ e^- e^+$ [4], which is highly suppressed within the standard model of elementary particles. This means that its observation would be a sign of new physics beyond the standard model. The Mu3e pixel detector should be able to cope with a muon decay rate of 10^9/s. It should have a momentum resolution of 0.5 MeV/c, a vertex resolution of 200 μm, and a time resolution of 20 ns. There will be 4-pixel layers with 80×80-μm² pixel size, 275 million pixels, and a total area of 1.9 m². The pixel detector thickness will only be 50 μm. The structure of the Mu3e detector is shown in Figure 2.17.

Within the Mu3e project, we have developed a 4×4-mm large detector prototype MuPix as a system of a chip in AMS 180-nm technology. This prototype is scalable to a size of 1×2 cm.

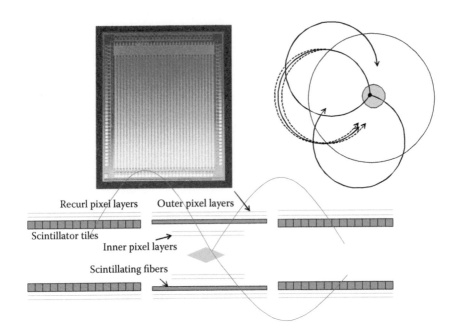

FIGURE 2.17   Pixel electronics.

2.9   MuPix

The MuPix detector is an HVCMOS detector with a 32 × 40-pixel matrix. The pixel size is 110 × 80 μm. The unit block of the chip is the pixel column. The pixel column contains 40 pixels (Figure 2.18) and two readout columns each with 20 readout cells and with 1 end-of-column (EoC) cell.

The chip contains 32-pixel columns. The pixels contain sensor diodes (nine deep n-well in p-substrate diodes, as shown in Figure 2.19). The central diode contains a charge-sensitive amplifier with a source follower (unity gain buffer) that is connected to its output, as shown in Figure 2.18. The electronics uses both types of transistors: NMOS and PMOS. PMOS transistors are placed directly inside the n-well. NMOS transistors are paced in a shallow p-well, which is inside the n-well. The schematics of the pixel electronics are shown in Figure 2.1.

Every pixel has its own readout cell that is placed in the lower part of the pixel column, i.e., on the chip periphery. The readout cell contains the second-stage amplifier, the comparator with 4-bit DAC, and 4-bit random access memory (RAM) to tune the offset and a digital part. The purpose of the second-stage amplifier is to additionally amplify the voltage signal that is generated by the pixel amplifier. The task of the comparator is to generate output signal (the hit signal) when a particle signal arrives. The threshold of the comparator should be higher than the noise level and lower than the lowest particle signal. The purposes of the digital part are (a) to generate and store a hit flag (a bit in a memory cell set/reset [SR] latch) after the comparator signal is issued, (b) to store the global time stamp when hit arrives, and (c) to allow the readout of the stored hit information (time stamp and cell address) on a priority

Amplifier

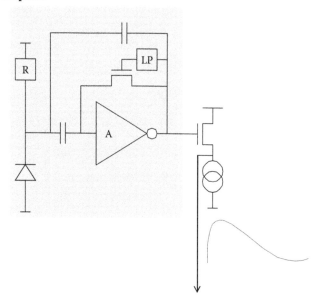

FIGURE 2.18   Layout of the pixel and the digital cell.

principle. (In the case of multiple hits, the upper cells send the information before the lower cells.) The layout of the readout cell is very dense; it occupies only 55 × 7 μm. The readout cells in a one-pixel column are arranged as two readout columns of 20 cells. As already mentioned in this section, there is an EoC cell that is attached to every readout column.

Let us now explain the priority-based readout. When the hit signal is issued by the comparator, precisely on its leading edge, the hit flag is set to one (see Figure 2.20).

In this moment, the current state of the 8-bit, gray-coded time stamp is stored into the RAM of the readout cell. By the issuing of the LdPix signal (external signal), all stored hits are confirmed. This means that the hit flag is stored into the second memory cell (confirmed hit). The confirmed hits cause a rippling down of the priority OR chain (HitIn/HitOut). When the rippling is finished, we will issue the RdPix signal (an external signal that can be blocked if the EoC cell is full). During the active level of RdPix, the readout cell with the confirmed hit that is the first in the priority chain puts its hit data onto the data bus. The hit data are the stored 8-bit time stamp and the content of the readout cell address read only memory (ROM) (readout row address). The RdPix signal also deletes the hit flag and the confirmed hit flag of the cell that is being read out. The state of the priority OR chain changes, and the next cell with the confirmed hit is selected for readout.

The purpose of the EoC cell (Figure 2.21) is to control the readout of the hit information from different readout columns, when the columns are connected together. In the case of multiple hits in different readout columns, the hits of the columns with higher priority will be read out first. The scheme is actually the same as the priority scheme in the columns that we explained above. Every EoC block has its own

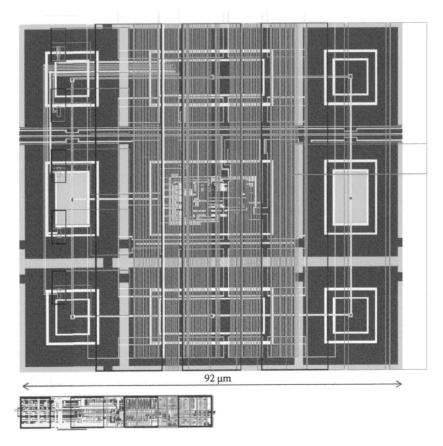

92 µm

FIGURE 2.19 Layout of the pixel and the digital cell.

memory cells for the hit information. When RdPix is high, the time stamp and the column address from the selected readout cell with a confirmed hit (the one with the highest priority) are copied into the EoC memory cells. Every EoC cell also contains a hit flag. This flag is set to one when RdPix is issued and if there are confirmed hits in the readout column. The EoC cells also have their priority OR chain (HitInCol/ HitOutCol) that is controlled by the outputs of the EoC hit flags. This chain controls the readout of the stored hit data. When another external-signal RdCol is active, the EoC cell with the stored hit that is the first in the OR chain sends the hit data onto the horizontal data bus. The hit data are 8-bit time stamp and the readout cell address. Additionally, the content of the EoC address ROM is transmitted too (column address). The RdCol signal also deletes the EoC hit flag of the EoC cell that is being read out. The state of the horizontal priority OR chain changes, and the next EoC cell with the hit is selected for readout. The RdCol cycles can be repeated until all the EoC cells are empty. Another possibility is to repeat the RdCol cycle n times, allowing the fact that some cells are not read out. After this, a new RdPix cycle is repeated to copy the next set of hits to the readout cells. If, by chance, one EoC cell

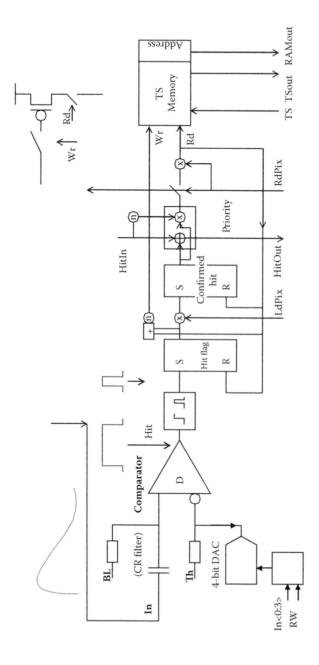

FIGURE 2.20 Readout cell. TS, time stamp.

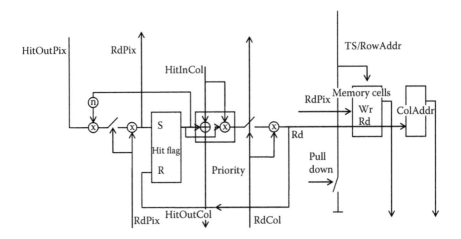

FIGURE 2.21 EoC cell.

is not emptied, but there is another hit in the attached column waiting for readout, this hit will not replace the one that is already stored in the EoC cell. No data transfer from the column to the EoC cell will happen because the RdPix is blocked when the EoC cell is not empty. After a certain number of RdPix and RdCol cycles (we can either repeat them until the priority out is zero and all hits are emptied or we repeat them a certain number of times), we will issue a new LdPix to confirm the new hits that are generated in the meanwhile.

REFERENCES

1. R. Turchetta et al. 2003. Monolithic active pixel sensors (MAPS) in a VLSI CMOS technology. *Nucl. Inst. Meth. B and A* 501, pp. 251–259.
2. C. Hu-Guo et al. 2009. CMOS pixel sensor development: A fast readout architecture with integrated zero suppression. *JINST*, 4, April.
3. M. Deveaux et al. 2007. Charge collection properties of monolithic active pixel sensors (MAPS) irradiated with non-ionising radiation. *Nucl. Inst. Meth. B and A* 583, pp. 134–138.
4. A. Blondel et al. 2012. Letter of Intent for an Experiment to Search for the Decay $\mu \rightarrow eee$. Available at http://www.psi.ch/mu3e/documents.

3 Analog Electronics for Radiation Detection

Juan A. Leñero-Bardallo
and Ángel Rodríguez-Vázquez

CONTENTS

3.1   Introduction 47
3.2   ADC Architectures for Image Sensors...49
    3.2.1   Pipeline ADCs...49
    3.2.2   Slope ADCs 49
    3.2.3   SAR ADCs 51
    3.2.4   Cyclic ADCs..52
    3.2.5 $\Sigma\Delta$ ADCs..53
3.3   ADC Topologies 53
    3.3.1   Global ADCs 53
    3.3.2   Column-Parallel ADCs..54
    3.3.3   Pixel-Level Analog-to-Digital Conversion 55
    3.3.4   Future Integrations..56
3.4   ADC Requirements for Image Sensors 58
    3.4.1   ADC Resolution...58
    3.4.2   Random Noise..60
    3.4.3   Area 61
    3.4.4   Speed 61
    3.4.5   Power Consumption...61
3.5   State of the Art 62
    3.5.1   ADC Efficiency and Design Considerations 62
    3.5.2   ADC Comparison 62
3.6   Qualification of Different ADC Architectures..66
3.7   Conclusions..66
References...67

3.1   INTRODUCTION

This chapter provides guidelines to choose and design analog-to-digital converters (ADCs) for image sensors. Results can be extrapolated to other kinds of radiation detectors on the focal plane. The main architectures usually employed to read out image sensor pixels' outputs are studied and compared. Modern ADC topologies are analyzed. The specific ADC requirements for image sensor applications are also described. The performance of relevant, recently published ADCs for image sensors is benchmarked. Finally, a discussion of the advantages and disadvantages of the different ADC architectures based on the state of the art is included. A specific figure of merit (FoM) to compare different ADCs for image sensors is proposed. ADCs are widely used for multiple purposes that are related to signal acquisition and data processing. Frame-based image sensors have traditionally included one ADC to read out their outputs and digitize them for further processing, representation, or storage. The image capture is a process with several steps. Image sensors transduce photons into an analog voltage signal that is proportional to the light intensity. Then, this analog output is converted into a digital signal, usually stored on a memory, and sent out for further representation. The quality of the analog-to-digital conversion affects the quality of the final image or frame. Our eyes are quite sensitive to the fixed-pattern noise (FPN) that is introduced during the analog-to-digital conversion. (Humans can detect a 0.5% change in mean intensity [1].) The number of bits of the ADC should be high enough to represent the output image with a number of gray levels that are similar to or higher than the number that our eyes can detect. The speed of the analog-to-digital conversion can limit the frame rate of our sensor. The noise introduced by the ADC should be controlled. Finally, the circuitry in charge of processing the output data flow provided by the converters has to be fast enough to avoid losing information.

Vision sensor designers are not usually experts in the design of ADCs. On the other hand, ADC designers are sometimes not aware of the special requirements that ADCs have to satisfy for image sensors. General-purpose ADCs are not suitable for commercial image sensors. Some of their specifications can be oversized leading to unnecessary power or area consumption. On the contrary, if they do not achieve some of the imager requirements for the signal digitization, they will degrade the image quality.

Traditionally, one global ADC was used to convert all the pixel outputs. ADC designers were not quite concerned about the power and area requirements. However, modern image sensors demand high frame rates and pixels with fine pitch. More than one ADCs are shared by different groups of pixels to increase the readout speed [2]. In this sense, it is crucial to decide how many ADCs are used to convert the pixel outputs, how they are distributed, how the pixel outputs are multiplexed, and the ADC requirements. The specifications of these ADCs can be very different from the converters that are dedicated to communications, for instance.

Emerging three-dimensional (3D) integration technologies also open new, enticing possibilities for ADC design. The conversion speed could be reduced by placing more ADCs in one tier [3]. ADCs could be shared by reduced groups of pixels with a fill factor that is close to 100%. The decision of how many ADCs are going to be placed and how the pixel outputs are going to be multiplexed is not a trivial matter.

The goal of this chapter is to guide image sensor designers to know the specific requirements that ADCs for image sensors must satisfy and how to distribute them and discuss the advantages and drawbacks of the different ADC architectures and topologies. This chapter is organized as follows: first, in Section 3.2, the most common ADC architectures for image sensors are described. Then, in Section 3.3, the main ADC topologies are analyzed. Next, in Section 3.4, the specific requirements that an ADC for image sensors must satisfy are discussed. Afterwards, in Section 3.5, the state of the art is analyzed, and the performance of relevant ADCs for image sensors is compared. Finally, in Section 3.6, the guidelines for a good choice of an ADC for image sensor applications are given, and the ADC performance of different architectures is compared.

3.2 ADC ARCHITECTURES FOR IMAGE SENSORS

There are four main ADC architectures that are usually employed for image sensor conversion: (1) pipeline, (2) slope (ramp ADCs), (3) cyclic, successive approximation registers (SAR), and (4) sigma delta ($\Delta\Sigma$). We also find hybrid ADCs that combine the advantages of different architectures. We describe in Sections 3.2.1 through 3.2.5 the principle of operation and advantages and disadvantages of each architecture.

3.2.1 PIPELINE ADCs

Pipeline ADCs [4] have a clocked topology with several operation stages. Each stage resolves a certain number of bits of the analog-to-digital conversion. The output of each stage is connected to the next. The m-most significant bits are computed by the first stage, and the rest of the bits are computed by the successive ones. There is the same delay between the output of each stage. Figure 3.1 shows the schematics of a pipeline ADC with m-bit conversion stages. Each stage resolves m-bits and is made up of one m-bit ADC and one m-bit digital-to-analog converter (DAC). Then, the remaining stage output is amplified with a gain $A = 2^m$ and goes to the next stage. Because the bits from each stage are determined at different clock cycles, all the bits corresponding to the same sample have to be time-aligned with shift registers. This imposes area requirements and circuit complexity that limits the usage of these converters with modern column-parallel implementations. They have been employed as global ADCs to read out imager pixel outputs [5]. Its main advantage is high conversion speed. Clock cycles of N_{bits} are required to digitize the input voltage. However, in every clock cycle, the same conversion bits are available from the previous input. The pixel output could be changed after reading the most significant bits of the first conversion stage, adding the possibility of implementing pipeline operation with the pixel readout. Thus, the equivalent pixel conversion speed would be only one clock cycle.

3.2.2 SLOPE ADCs

Slope ADCs (also known as ramp ADCs) are the most extended converters for image sensors with column-parallel readout topology. Their architecture is rather simple (see Figure 3.2). They have a ramp generator, a comparator, a counter, and an output memory. Initially, the ramp generator and the counter are reset. When the conversion starts, the counter and the ramp generator are activated. If the slope voltage exceeds the input voltage, the counter is stopped, and its outputs are latched. Their

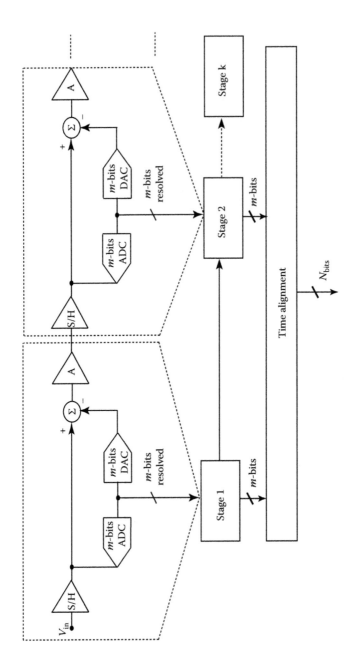

FIGURE 3.1 Pipeline ADC block diagram.

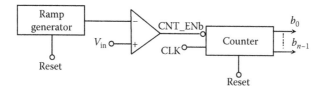

FIGURE 3.2 Ramp ADC block diagram and principle of operation. CLK, counter clock; CNT_Enb, counter outputs enable.

advantages are simplicity, low area requirements (they can be lay out with a fine pixel pitch), and acceptable FPN. Their low area requirements are a great advantage for modern image sensors with a large number of pixels. They can even be embedded inside individual pixels [6] leading to high-speed image sensors.

The main disadvantage of slope ADCs is their conversion speed since $2^{N_{bits}}$ of clock cycles are required to carry out a conversion. Thus, they are more suitable for ADCs with a low bit resolution. To overcome this limitation, several authors propose multiramp ADCs. The input signal is compared to a fast slope to determine the coarse bits of the input signal. Then, a slope with an offset that corresponds to the course of the input signal is activated to determine the less significant bits of the input signal.

There are also some other issues with this architecture: (a) misalignment between coarse and fine conversion ramps and (b) circuit complexity that grows exponentially with the number of coarse-bit conversion. FPN related to ramp ADCs is mainly due to the mismatch of the ramp generators. Ramp generators commonly use capacitors. To improve this limitation, the ramp generators' capacitance is increased. This imposes area requirements to the ADC layout.

3.2.3 SAR ADCs

SAR ADCs overcome the speed limitations of ramp ADCs. SAR ADCs require only N_{bits} clock cycles to perform an ADC conversion. The main limitations of SAR ADCs are area and power consumption. Their principle of operation is as follows (see Figure 3.3): initially, the input analog signal is sampled and held. Then, it is compared to a reference voltage value. The comparison result sets the value of the most significant bit (MSB). Then, the reference voltage is changed, and the comparison is repeated N_{bits}—one time. The control signal end of cell indicates the end of the analog-to-digital conversion. A DAC and digital logic are employed to set the voltage reference values. This imposes expensive area requirements that makes it challenging to implement a SAR for each column. To amend this limitation, it would be possible to share an ADC by a group of pixels. However, in that case, the readout speed requirements would increase. Thus, there is a trade-off between area requirements and speed. Another disadvantage of this kind of converters is the mismatch that is introduced by the DAC. Some authors [7] try to overcome it by calibrating the reference voltage of different DACs. The penalties are more area consumption and circuit complexity. In any case, the DAC design should take into account mismatch considerations to reduce FPN.

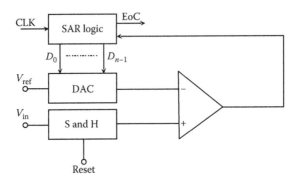

FIGURE 3.3   SAR ADC block diagram.

3.2.4 CYCLIC ADCs

The principle of operation of cyclic ADCs is similar to SAR ADCs. They only need N_{bits} clock cycles to operate. The main difference is that the reference voltage, V_{ref}, does not need to be generated during each iteration. Hence, a DAC is not necessary. They employ switched-capacitor (SC) amplifiers, which amplify the difference between the input signal and the reference voltage (see Figure 3.4).

The input signal is changed after each operation cycle. It will be equal to the remaining voltage at the output of the amplifier. After each operation cycle, the remaining input signal is compared to V_{ref} to generate the next MSB. Depending on the value of the MSB, either V_{ref} or zero will be subtracted from the remaining input voltage in the next cycle.

The voltage at the output of the amplifier V_{resout} can be expressed as

$$V_{resout}(k) = A \cdot V_{resin}(k) - d_k \cdot V_{ref} \tag{3.1}$$

where k is the number of the conversion cycle.

Variations of the amplifiers gain can lead to high FPN that can be unacceptable for high-quality image sensors. This imposes a good capacitor matching (more area requirements) and higher currents for the amplifier biasing (more power consumption). FPN and area are the main limitation of cyclic converters for image sensors.

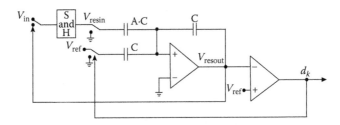

FIGURE 3.4   Cyclic ADC block diagram.

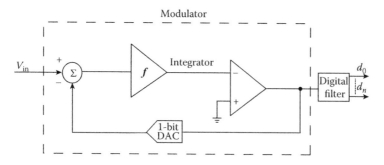

FIGURE 3.5 ΣΔ ADC block diagram.

3.2.5 ΣΔ ADCs

ΣΔ ADCs (see Figure 3.5) have traditionally been used as global ADCs with imagers with low speed and low FPN requirements. They consist of a ΣΔ modulator followed by a digital decimation filter. The modulator has an integrator and a comparator with a feedback loop that contains a 1-bit DAC. The DAC just connects the comparator input to a positive or a negative reference voltage. A clock is necessary to provide the correct timing for the modulator and the digital filter. These kinds of converters suppress the input noise by oversampling. The voltage variations of the input signal are tracked quickly by ΣΔ converters. However, for image sensors with column-parallel readout, it is not necessary to track fast variations of the input signal that can be considered static during the readout process. For this reason, some authors design incremental ΣΔ converters that are optimized to read quasi-static DC signals and negligible offset [8–10].

Their main advantages are low FPN, good resolution, and speed. Their disadvantages are the area requirements, circuit complexity for column-parallel implementation, and their continuous operation mode that can lead to an unnecessary power consumption when they are employed to read out static image sensors' outputs. As it will be discussed in Section 3.6, some of these drawbacks can be solved by adapting the design of ΣΔ converters to the specific requirements of image sensors leading to a very competitive emerging generation of converters.

3.3 ADC TOPOLOGIES

In this section, we are going to discuss the different topologies [2] to digitize pixel outputs. We will study how the conversion speed can be increased by adding more than one ADC per pixel array.

3.3.1 GLOBAL ADCs

The traditional approach was to use a single ADC per sensor. All the pixel outputs were time-multiplexed to the input of a global ADC, as shown in Figure 3.6a,

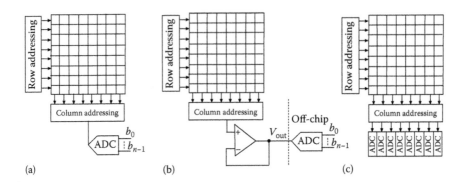

FIGURE 3.6   (a) Readout scheme with a global ADC. (b) Readout scheme with global internal buffer for an external ADC. (c) ADCs in column-parallel distribution.

to provide a digital output. The analog-to-digital conversion frequency of an entire frame is given by the following equation [2]:

$$f_{conv} = \left[M \cdot N \cdot \left(\tau_{ADC} + \tau_{RO} \cdot \frac{n_{bits}}{n_{parallel}} \right) \right]^{-1} \qquad (3.2)$$

where M and N are the number of pixels per row and column, respectively, τ_{ADC} is the ADC sampling time ($\tau_{ADC} = 1/f_s$), τ_{RO} is the time that is required to send 1 bit out of the chip, which depends on the master clock frequency, n_{bits} is the number of bits per sample, and $n_{parallel}$ is the number of parallel digital outputs. The ADC was usually integrated on the chip to maximize the frame rate and reduce the FPN. Global ADCs were usually designed to maximize the readout speed. Designers were not very concerned about the area requirements or the power consumption of the ADC. The pixel array area and power consumption were usually much higher than the ADC area and power consumption.

Another approach with a global ADC was to provide an analog readout. The ADC was placed off-chip (see Figure 3.6b). In this case, pixel outputs were first time-multiplexed and then connected to the input of a global internal analog buffer. Analog intermediate memories could also be used to increase the frame rate of the sensor [11]. The internal buffer should be fast enough to guarantee the desired frame rate. It should also be able to provide enough output current to charge or discharge on time the high capacitive load that is introduced by the output pad and the long metal lines.

3.3.2   COLUMN-PARALLEL ADCS

The demand of commercial sensors with high resolution and high frame rates requires to significantly reduce the time that is dedicated to read out the pixel outputs. Under these requirements, global ADCs are not practical and have been deprecated during the previous years. Obviously, by increasing the number of ADCs working in parallel with different groups of pixels, the frame readout time will be reduced.

For simplicity, by placing one ADC per column (see Figure 3.6c), all the pixels of one row can be sampled and read out simultaneously. This approach has a twofold benefit: the frame readout speed is increased significantly, and the requirements for the conversion time are less restrictive. The frequency of the frame analog-to-digital conversion is almost M times faster:

$$f_{conv} = \left[M \cdot N \cdot \left(\frac{\tau_{ADC}}{M} + \tau_{RO} \cdot \frac{n_{bits}}{n_{parallel}} \right) \right]^{-1} \tag{3.3}$$

The main challenges are the area requirements for the ADC. The ADC width should be lower than the pixel pitch. The power consumption should be low because the number of ADCs will be increased significantly. Some authors [12] place one ADC per column alternatively on the top and bottom of the array. Thus, the ADC pitch can be doubled relaxing the ADC area design requirements and achieving the same readout speed than with the column-parallel implementation of Figure 3.6.

It would be even possible to divide the sensor into two blocks and place an ADC per column at the top and bottom of the sensor. The two blocks could be read out in parallel employing a correlated double sampling approach, almost doubling the conversion frequency. This technique is more attractive with pixels with large pitch, typically biomedical arrays with low resolution, where the ADC area and the power requirements are less exigent.

Another issue to take into account with multiple ADCs per sensor is the output data flow. We should have circuitry to store and send out all the ADC outputs. External devices connected to our sensor should also be able to digest the output bit stream.

3.3.3 Pixel-Level Analog-to-Digital Conversion

Some authors [6,13,14] propose digital pixel sensors (DPSs) with integrated ADCs into each pixel. The DPS performance takes advantage of the CMOS scaling-down properties. The idea is simple: each pixel has a dedicated ADC and provides an independent digital output, considerably increasing the readout speed. In this case, the conversion frequency is given by

$$f_{conv} = \left[\tau_{ADC} + \tau_{RO} \cdot M \cdot N \cdot \frac{n_{bits}}{n_{parallel}} \right]^{-1} \tag{3.4}$$

Although the conversion speed is higher, there are important drawbacks. This method reduces the pixels' fill factor. It also requires massive parallel circuitry that is capable of reading large amounts of data.

Figure 3.7 compares f_{conv} for the three different readout topologies for different array sizes $(M \cdot N)$. We assume for the computation that, $M = N$, $\tau_{ADC} = 1$ μs, $\tau_{RO} = 10$ ns, and $n_{bits} = n_{parallel} = 12$. These values correspond to standard modern ADC features (see Section 3.5 for more details). The value f_{conv} is always higher using multiple ADCs. For arrays with a low number of pixels, the DPS topologies achieve the

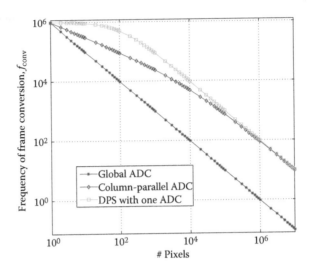

FIGURE 3.7 Frequency of conversion for different ADC topologies: (*) global ADCs, (◊) column-parallel ADCs, and (□) pixel-level ADCs.

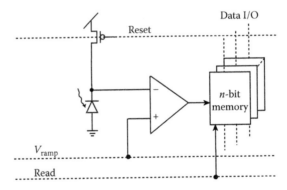

FIGURE 3.8 Diagram of pixel with an internal ramp ADC and n-bit digital memory.

highest speed. However, for large-pixel arrays, this topology offers the same performance than the column-parallel topology. For this reason, DPSs are usually aimed for high-speed image sensor applications with low-resolution arrays.

Figure 3.8 shows a DPS with an integrated n-bit analog converter. The pixel has an n-bit ramp converter and a memory to save its outputs. To reduce the pixel area, some circuitry of the ADCs typically can be shared by the different pixels. In this example, the ramp generator and the counters are shared by all the pixels. Each pixel requires a comparator and a memory to perform the pixel-level analog-to-digital conversion.

3.3.4 FUTURE INTEGRATIONS

The development of 3D technologies with stacked and interconnected dies opens new possibilities to increase the readout speed without reducing the fill factor. One entire die (tier) could be dedicated for the ADC design. Ideally, the photoactive area could be placed on the top level, and an ADC per pixel could be placed below it (see Figure 3.9). Kiyoyama et al. recently reported a very interesting ADC for 3D integration [3]. Each pixel block was connected to a dedicated ADC on the bottom tier with a pitch of 100×100 µm². However, one dedicated ADC per pixel is not feasible with photodiodes with a fine pitch. In this case, it would not be possible to fit an ADC or a processing circuitry below the photodiodes. Thus, the pitch of the processing circuitry will be higher than the photodiode pitch. This implies design challenges to face: (a) groups of photodiodes should share processing circuitry, and (b) they should be multiplexed. Figure 3.10 illustrates a possible implementation with multidistributed processing units (PUs). The pitch of the PUs is bigger than the photodiode pitch. In such topology, f_{conv} is given as

$$f_{conv} = \left[M \cdot N \cdot \left(\frac{\tau_{ADC}}{N_{adc}} + \tau_{RO} \cdot \frac{n_{bits}}{n_{parallel}} \right) \right]^{-1} \qquad (3.5)$$

where N_{adc} is the number of ADCs on the bottom tier.

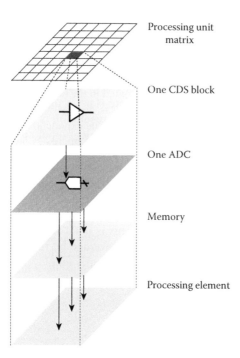

Processing unit matrix

One CDS block

One ADC

Memory

Processing element

FIGURE 3.9 Diagram of a 3D stacked system with dedicated ADC, memory, and processing elements per pixel. (Adapted from K. Kiyoyama et al., A very low area ADC for 3D stacked CMOS image processing system, *IEEE International 3D Systems Integration Conference, 3DIC*, pp. 1–4, 2011.)

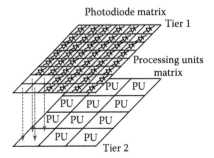

FIGURE 3.10 Possible future 3D integration. Top layer is dedicated to photodiodes. Bottom layer is dedicated to PUs.

3.4 ADC REQUIREMENTS FOR IMAGE SENSORS

In this section, we are going to discuss the specific ADC requirements for image sensors. ADC designers are usually concerned about resolution, speed, and power. As we will discuss, an ADC for image sensors does not need to satisfy strict requirements, but it is desirable to optimize its design to save area and power consumption while maximizing the conversion speed.

3.4.1 ADC RESOLUTION

Figure 3.11 displays an image that is digitized with different numbers of bits. With 4 bits, we clearly notice a loss of image quality for using an insufficient number of bits. With six or more bits, there is not a significant improvement on the image quality. However, a much better resolution is needed to properly capture dark images, when the converter's equivalent input noise is similar to pixel noise. We can wonder how many gray levels our eyes can visualize [15]. There is no clear answer, but several studies have converged that they are roughly 450 under certain illumination conditions. The Digital Imaging and Communications in Medicine (DICOM) standard sets a maximum of 450 gray levels for display representation, for instance. This means that 9-bit resolution is enough for an ADC conversion of the pixel outputs. This is not a very restrictive ADC resolution requirement, taking into account the average bit resolution of the currently reported ADCs. ADCs for image sensors usually employ 10 or more bits.

FIGURE 3.11 Number of bits' effect on image quality. From left to right: 4-, 6-, and 8-bit precision.

The photon shot noise of image sensors can even be exploited to relax the requirements of quantization noise that is introduced by ADCs [16]. Photon shot noise is due to the temporal variations of the number of incident photons. It becomes the dominant noise source with high levels of illumination. Figure 3.12 shows the dependence of a sensor output and various noise sources with illumination. The output has a linear dependence with the illumination level until it saturates at level N_{sat}, which corresponds to the full well capacity of the sensor. Almost all the noise sources are independent of illumination ($1/f$ noise, thermal, etc.) and set a noise floor that limits the sensor's dynamic range of operation. However, the photon shot noise depends on the input signal level N_{sig} (number of electrons generated by the photodiode during the integration period) as follows:

$$e_{phs}(N_{sig}) = \sqrt{N_{sig}} \tag{3.6}$$

Photon noise is dominant with bright illumination. We can see that the ADC in this region has less quantization noise than required. Hence, it would be possible to relax the ADC requirements that are related to resolution to reduce the power consumption or increase the speed, for instance.

We can define a quality parameter r that is given by the ratio of the quantization noise and the photon noise due to the signal as

$$r = \frac{e_{qns}(k)}{e_{phs}(N_{sig})} \tag{3.7}$$

The quantization noise $e_{qns}(k)$ depends on the quantization step size e_{lsb} and the number of quantization steps k, as follows:

$$e_{qns}(k) = k \cdot \frac{e_{lsb}}{\sqrt{12}} \tag{3.8}$$

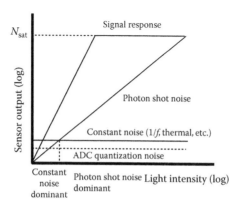

FIGURE 3.12 Dependence of the sensor output and the main noise sources with illumination. (Adapted from M. F. Snoeij et al., *IEEE Journal of Solid-State Circuits*, vol. 42, no. 12, 2007.)

For optimal ADC design, the input range of the ADC should be matched to the maximum output swing of the sensor N_{sat}:

$$N_{sat} = 2^n \cdot e_{lsb} \tag{3.9}$$

And, combining Equations 3.6 through 3.9, we can compute the input signal levels at which we can double the quantization noise, i.e., to reduce the ADC resolution to 1 bit, maintaining the ratio between the quantization and the photon shot noise constant:

$$N_{sig} = \left(\frac{k \cdot N_{sat}}{2^n \cdot r \cdot \sqrt{12}} \right)^2 \tag{3.10}$$

Figure 3.13 illustrates how the photon shot noise can be exploited to increase the quantization levels. An elegant design that exploits this property with multiramp ADCs was described by Snoeij et al. [15].

3.4.2 RANDOM NOISE

Another feature to take into account is the random noise that is introduced by the converter. It can be expressed in volts, but image sensor designers usually express it in electrons, which is related to the full well capacity of the vision sensor. Random noise provokes column FPN. This is a mismatch that is introduced by the different column ADCs in a column-parallel readout topology. If the different ADCs connected to each column introduce mismatch, we will perceive column variations of the gray levels when the array is illuminated with uniform illumination. As mentioned in Section 3.1, humans can detect a 0.5% change in mean intensity [1]. Therefore, the FPN introduced by an ADC is a very important parameter to take into account for high-quality image sensors. Figure 3.14 illustrates the effect of FPN in a column-parallel readout topology. One captured image with three different levels of FPN is shown. On the right, the

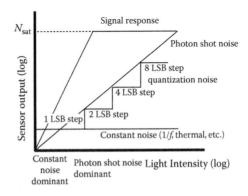

FIGURE 3.13 Illustration of how shot noise can be exploited to relax the ADC quantization noise specifications. (Adapted from M. F. Snoeij et al., *IEEE Journal of Solid-State Circuits*, vol. 42, no. 12, 2007.)

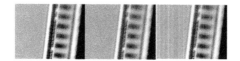

FIGURE 3.14   FPN's effect on image quality.

degradation provoked by FPN on the image is clearly displayed. We observe pixel columns with different gray levels in regions with uniform illumination. This undesirable effect can be improved if the ADCs do not always read the output of the same column in a column-parallel implementation. If the ADC reads out a column that is shifted every time that a different row is selected, the column FPN will be improved.

3.4.3   AREA

The area requirements are restrictive for image sensors. If we employ one ADC per column, the ADC pitch must be equal to or lower than the pixel pitch. This limits in many cases the nature of the ADC that we can use. Nowadays, we can find very fine pixel pitches below 1 µm in commercial image sensors. The ADC layout can be very tricky and complicated. Capacitors require a large area to be fitted in a rectangle width that is lower than the pixel pitch. Metal-insulator-metal (MiM), metal-oxide-metal (MoM), and polycapacitors are commonly used to implement SC circuits. Depending on the technology, we can find design rules that limit the pixel pitch.

3.4.4   SPEED

The speed of the analog-to-digital conversion is an important parameter to consider. The ADC sampling frequency, $f_s = 1/\tau_{ADC}$, should be higher than

$$f_s \gg \frac{1}{\dfrac{N_{ADC}}{FR \cdot N_{pixels}} - \dfrac{\tau_{RO} \cdot N_{ADC}}{n_{bits}}} \tag{3.11}$$

where FR is the target frame rate, N_{ADC} is the number of ADCs that are shared by all the pixels, τ_{RO} is the amount of time to send 1 bit out the chip, and $n_{parallel}$ is the number of digital outputs. Nowadays, for high-resolution image sensors, usually, $N_{ADC} > 1$. The maximum speed can be achieved when an ADC per pixel is implemented. In that case, all the pixels of the array can be read out simultaneously. Thus, there is a trade-off between readout speed, pixel complexity, and output sensor throughput.

3.4.5   POWER CONSUMPTION

Since most image sensors are used in mobile devices, power consumption has become an important drawback. Nowadays, the dominant component of power consumption in CMOS image sensors with column-parallel ADCs is analog-to-digital conversion followed by output readout [17,18]. Thus, there is a trade-off between conversion speed and consumption. Power consumption in CMOS image sensors increases at least linearly in resolution and frame rate.

3.5   STATE OF THE ART

We summarize in Table 3.1 some relevant ADCs that were recently reported in the *IEEE Journal of Solid-State Circuits, IEEE International Solid-State Circuits Conference, VLSI Symposium, IEEE Transactions on Electronic Devices, IEEE Sensors Journal*, and other relevant journals. All of them are targeted for image sensors with parallel readout. They have been tested with pixel arrays of quarter video graphics array (QVGA) resolution or higher. For each converter, we indicate the resolution, the minimum pixel pitch that can be read out, the power consumption, the random noise that is introduced by the converter, the conversion frequency, and the clock frequency of operation.

3.5.1   ADC Efficiency and Design Considerations

Looking at the ADCs for image sensors reported in Table 3.1, we can wonder how efficient they are in terms of energy consumption. There is a theoretic energy that is bound for the sampling energy that is imposed by thermal noise [19]:

$$E_{min} = 8 \cdot k \cdot T \cdot SNR = 8 \cdot k \cdot T \cdot 1.5 \cdot 2^{N_{bits}} \qquad (3.12)$$

In Figure 3.15, we have plotted the energy per sample of each converter that is normalized by E_{min}. Murmann [20] established a limit of $E_s/E_{min} = 100$ for the most efficient current high signal/noise ratio converters, for which energy consumption is just limited by thermal noise (when signal-to-noise and distortion ratio (SINADR) > 60 dB). Looking at Figure 3.15, we see that the most energy-efficient converters for image sensors are still two orders of magnitude above the Murmann's energy limit for modern ADCs. SAR and cyclic converters are more efficient. Slope converters are less efficient and appear on the top of the plot far away from the Murmann's bound. Therefore, the choice of the ADC architecture has an influence on the energy efficiency that should be taken into account.

Figure 3.16 displays the dependency between the ADC resolution and the energy consumption per sample $E_s = P/f_s$. There is an exponential dependency between them. Increasing the bit resolution provokes an exponential increment of the energy consumption. We also have to consider that the total ADC power consumption will be multiplied by the number of columns in a column-parallel implementation. As it was discussed in Section 3.4.1, increasing the ADC resolution does not necessarily improve the image quality beyond nine effective bits. Therefore, limiting the number of effective bits to 10 will save a lot of power.

3.5.2   ADC Comparison

The imager's requirements can be very different depending on the application: surveillance, industrial inspection, video, microscopy, photography, etc. Some applications

TABLE 3.1

Relevant ADCs for Image Sensors Recently Published

Ref.	ADC Type	Resolution	Random Noise	Pixel Pitch	Power	f_s/CLK
[16]	Multiple slope	10b	Adaptive to light intensity	7.5 μm	130 μW	0.8 MS/s/ 20 MHz
[17]	Cyclic	13b	2.5 e⁻ (153 μV)	5.6 μm	300 μW	19.2 MS/s/ 250 MHz
[18]	ΣΔ	12b	1.9 e⁻	2.25 μm	148 μW	0.14 MS/s/ 48 MHz
[21]	Single slope	12b	1.1 e⁻ (121 μV)	1.4 μm	100 μW	31.7 KS/s/ 130 MHz
[22]	Cyclic with folding integration	13–19b	1.2 e⁻ (80 μV)	7.5 μm	436 μW	1.5–2.3 MS/s/ 30 MHz
[23]	SAR	14b	2.8 e⁻ (100 μV)	4.2 μm	41 μW (DC)	3.5 MS/s/ 49.5 MHz
[24]	Single slope	13b	ND	2.5 μm	ND	6.6 KS/s/ 54 MHz
[25]	Single slope	12b	ND	2.97 μm	ND	0.13 MS s/ 2.37 GHz
[26]	ΣΔ + cyclic	14b	ND	3.9 μm	ND	0.125 MS/s/ 5 MHz
[27]	Single/ multiple slope	10b	Adaptive to light intensity	10 μm	ND	0.43–0.047 MS/s/ 49 MHz
[28]	Cyclic	17b	1.17 e⁻ (21 μV)	7.1 μm	ND	31.15 KS/s/ 1.37 MHz
[29]	SAR	10b	ND	2.25 μm	41 μW	0.56 MS/s/ 7 MHz
[30]	SAR	11b	527 μV	7 μm	209 μW	0.83 MS/s/ 10.8 MHz
[31]	SAR	10b	240 μV	10 μm	35.46 μW	240 KS/s/ 245.76 KHz
[12]	Cyclic	12b	1800 μV	10 μm	430 μW	0.17 MS/s/ 2 MHz
[32]	SS/SAR	11b	1200 μV	3.5 μm	7 μW	0.83 MS/s/ 2 MHz
[33]	SAR	9b	5300 μV	7.4 μm	1.37 μW	33.3 KS/s/ 2.06 MHz
[3]	SAR	9b	335 μV	100 μm (3D integration)	381 μW	4.4 MS/s/ 10 MHz
[9]	ΣΔ	13.5b	70 μV	10 μm (3D integration)	200 μW	46.7 KS/s/ 20 MHz

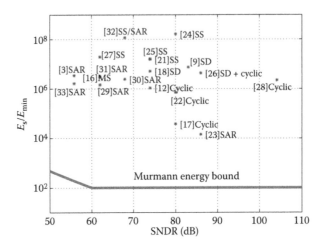

FIGURE 3.15   ADC conversion energy normalized to E_{min} and Murmann's energy bound.

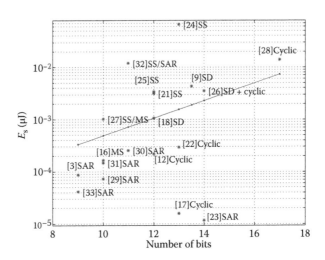

FIGURE 3.16   Energy per sample consumption versus ADC bit resolution. We have plotted in red an exponential data fitting.

require not only low image quality but also low power consumption or high-speed operation (surveillance or industrial inspection), for instance. Others require high image quality (fine pixel pitch) and are less restrictive with operation speed or power consumption, i.e., high-quality imaging. Therefore, ADCs for image sensors should be compared taking into account the scope of application of the imager whose outputs have to be converted. There are specific FoMs to compare ADCs (*International Solid-State Circuits Conference's* FoM, Walden's FoM, Emmert's FoM [34], Jonsson's FoM [35], etc.). However, there are no FoMs for ADCs for image sensors.

In order to compare the performance of the different ADCs for image sensors, we define the following FoM:

$$\text{FoM} = \frac{N_s \cdot \text{pitch}^2 \cdot P}{2^{N_{bits}} \cdot f_s}$$

(3.13)

The FoM was written in the *lower-is-better* form to simplify comparison. It rewards the number of bits, N_{bits}, and the speed sampling frequency, f_s. It penalizes the random noise, N_s, of the ADC (expressed in V_{rms}), the power consumption, P, and the minimum pixel width that can be read out (pitch). As it was discussed in Section 3.4.3, maybe the major design constraint with modern ADC topologies for image sensors is the ADC width (pitch). The required value of the sampling frequency, f_s, should be calculated taking into account the ADC topology, as explained in Section 3.3. The power consumption should be multiplied by the total number of ADCs.

We have compared the column-parallel ADCs of Table 3.1 using the specific FoM that was just defined. Table 3.2 summarizes the FoM values of all the sensors that were reported in Table 3.1. To make the comparison, we assigned the worst reported parameters to the nondisclosed (ND) values of the ADCs that were referenced in Table 3.1.

TABLE 3.2
FoM Values for the ADCs of Table 3.1

Reference	ADC Type	FoM ($\mu J \mu V_{rms} \mu m^2$)
[23]	SAR	1.26 e − 6
[17]	Cyclic	9.15 e − 6
[28]	Cyclic	1.13 e − 4
[22]	Cyclic with folding integration	1.59 e − 4
[21]	Single slope	1.82 e − 4
[29]	SAR	1.9 e − 4
[18]	$\Sigma\Delta$	6.88 e − 4
[26]	$\Sigma\Delta$ + cyclic	0.0017
[9]	$\Sigma\Delta$	0.0026
[30]	SAR	0.0032
[31]	SAR	0.0035
[25]	Single slope	0.0038
[15]	Multiple slope	0.0047
[33]	SAR	0.0220
[24]	Single slope	0.0266
[27]	Single/multiple slope	0.0522
[32]	SS/SAR	0.10
[12]	Cyclic	0.15
[3]	SAR	0.5666

According to this FoM, ramp ADCs [21,24,25] perform worse than SAR or cyclic converters. They are not suitable for high-speed image sensors. For instance, the ramp ADC reported by [25] can operate with conversion rates up to 0.13 MHz, but it needs very fast internal clocks at 2.37 GHz. However, they can be integrated with very fine pitches, and this makes them popular. The SAR ADCs [23,29] offer good performance, with pitches of 4.2 and 2.25 μm, respectively. ΣΔ ADC [18,26] offers a good compromise between all the parameters that we are considering. Finally, cyclic ADCs [17,22,28] score very well with our FoM, but they are limited to pixels with higher pitches than other converters (above 5 μm).

3.6 QUALIFICATION OF DIFFERENT ADC ARCHITECTURES

Although SAR ADCs require a DAC per column, whose area is large for consumer electronics with a fine pixel pitch, they have been adapted and optimized for column-parallel topologies offering very competitive solutions [23,29], according to our FoM. Modern implementations achieve a fine pitch, a good resolution, and a low power consumption.

Cyclic topologies also perform very well according to our FoM. However, cyclic ADCs are less common in commercial column-parallel implementations [17,28]. They are limited to pixels with high pitch. They also require high power consumption. We find interesting and competitive designs in the literature with pixel pitches over 5 μm [17,22]. There are reported hybrid architectures with cyclic converters that exploit some of their advantages [26].

Ramp ADCs score worse with our FoM. However, they can be integrated with very fine pitches and are very frequently used with image sensors [21,24,25]. Their advantages are the low circuit complexity and the low area requirements, with an acceptable noise for image sensors with column-parallel topologies. Their drawback is the low speed because they need faster clocks. According to the most recent, relevant publications in the field of image sensors, ramp ADCs are the most used for their integration in column-parallel structures with a fine pitch. Although they have lower speed than other architectures, their conversion time is enough for the majority of low-speed imaging applications.

ΣΔ converters have been traditionally deprecated and less investigated for column-parallel readout architectures due to their circuit complexity and high area requirements. They usually have been designed for low-noise image sensors, offering lower noise levels and higher speed than ramp converters. Their complexity due to the ΣΔ modulator and the following decimation filters makes it challenging to integrate them with low pitches. However, we find in the literature interesting designs like [18], with a ΣΔ-reduced modulator that is implemented with inverted-based SC circuits, or [26], with a hybrid ΣΔ and cyclic ADC with very low FPN. Specific incremental ΣΔ converters for quasi-static or DC measurements (e.g., imaging or temperature sensors) have also been reported [8–10,36].

3.7 CONCLUSIONS

In this chapter, the advantages and disadvantages of the main ADC architectures for image sensors have been discussed. The special requirements that ADC for image sensors must satisfy have been defined. We have also presented and compared the frequency of conversion for the main ADC topologies that are suitable for modern commercial sensors. Relevant and recent publications in analog-to-digital converters for image sensors have been reviewed and compared. Moreover, we have defined a specific FoM to compare the different ADCs for image sensors. Finally, the chapter gives guidelines for a good choice of an ADC for image sensors depending on the desired frame rate, the pixel pitch, the noise level that is introduced by the converter, and the target power consumption. Chapter results can be extrapolated to any kind of radiation detector in the focal plane that is made up of a pixel array.

REFERENCES

1. H. R. Blackwell. 1946. Contrast threshold of the human eye. *Journal of the Optical Society of America*, vol. 36, no. 11, pp. 624–643, November.
2. M. El-Desouki, M. J. Deen, Q. Fang, L. Liu, F. Tse, and D. Armstrong. 2009. CMOS image sensors for high speed applications. *Sensors*, vol. 9, no. 1, pp. 430–444.
3. K. Kiyoyama, K. Lee, H. Fukushima, T., Naganuma, Kobayashi, H., T. Tanaka, and M. Koyanagi. 2011. A very low area ADC for 3D stacked CMOS image processing system. *IEEE International 3D Systems Integration Conference, 3DIC*, pp. 1–4, January.
4. J. Ruiz-Amaya, M. Delgado-Restituto, and A. Rodríguez-Vázquez. 2011. Device-level modeling and synthesis of high-performance pipeline ADCs. New York: Springer.
5. S. Hamami, L. Fleshel, and O. Yadid-Pecht. 2006. CMOS image sensor employing 3.3 V 12 bit 6.3 MS/s pipelined ADC. *Sensors and Actuators*, vol. 135, no. 1, pp. 119–125, July.
6. S. Kleinfelder, S. Lim, X. Liu, and A. E. Gamal. 2001. A 10,000 frames/s CMOS digital pixel sensor. *IEEE Journal of Solid-State Circuits*, vol. 36, no. 12, pp. 2049–2059, December.
7. A. I. Krymski, N. E. Bock, N. Tu, D. V. Blerkom, and E. R. Fossum. 2003. A high-speed, 240-frames/s, 4.1 Mpixel CMOS sensor. *IEEE Transactions on Electron Devices*, vol. 50, no. 1, pp. 130–135, January.
8. J. Robert, G. C. Temes, V. Valencic, R. Dessoulavy, and P. Deval. 1987. A 16-bit low-voltage CMOS A/D converter. *IEEE Journal of Solid-State Circuits*, vol. 22, no. 2, pp. 157–163.
9. A. Xhakoni, H. Le-Thai, and G. Gielen. 2014. A low-noise high-frame-rate 1D-decoding readout architecture for stacked image sensors. *IEEE Sensors Journal*, vol. 14, no. 6, pp. 1966–1973, June.
10. J. Markus, J. Silva, and G. Temes. 2004. Theory and applications of incremental delta-sigma converters. *IEEE Transactions on Circuits and Systems I*, vol. 51, no. 4, pp. 678–690, April.
11. T. Sugiyama, S. Yoshimura, R. Suzuki, and H. Sumi. 2002. A 1/4-inch QVGA color imaging and 3D sensing CMOS sensor with analog frame memory. *IEEE International Solid-State Circuits Conference, Digest of Technical Papers*, San Francisco. pp. 434–435, February.
12. M. Furuta, Y. Nishikawa, T. Inoue, and S. Kawahito. 2007. A high-speed, high-sensitivity digital CMOS image sensor with a global shutter and 12-bit column-parallel cyclic A/D converters. *IEEE Journal of Solid-State Circuits*, vol. 43, no. 4, pp. 766–774, April.
13. D. X. D. Yang, B. Fowler, and A. E. Gamal. 1999. A Nyquist-rate pixel-level ADC for CMOS image sensors. *IEEE Journal of Solid-State Circuits*, vol. 34, no. 3, pp. 237–240, March.

14. A. Kitchen, A. Bermak, and A. Bouzerdoum. 2005. A digital pixel sensor array with programmable dynamic range. *IEEE Transactions on Electron Devices*, vol. 52, no. 12, pp. 2591–2601, December.

15. A. Torralba. 2009. How many pixels make an image? *Visual Neuroscience*, vol. 26, pp. 123–131, February.

16. M. F. Snoeij, A. J. P. Theuwissen, K. A. A. Makinwa, and J. H. Huijsing. 2007. Multiple-ramp column-parallel ADC architectures for CMOS image sensors. *IEEE Journal of Solid-State Circuits*, vol. 42, no. 12, pp. 2968–2977.

17. J.-H. Park, S. Aoyama, T. Watanabe, K. Isobe, and S. Kawahito. 2009. A high-speed low-noise CMOS image sensor with 13-b column-parallel single-ended cyclic ADCs. *IEEE Transactions on Electron Devices*, vol. 56, no. 11, pp. 2414–2422.

18. Y. Chae et al. 2010. A 2.1 Mpixel 120frame/s CMOS image sensor with column-parallel $\Sigma\Delta$ ADC architecture. *International Solid-State Circuits Conference, ISSCC*, vol. 46, no. 1, pp. 236–247.

19. E. A. Vittoz and Y. P. Tsividis. 2002. Frequency-dynamic range-power. In *Trade-Offs in Analog Circuit Design*, C. Toumazou, G. Moschytz, B. Gilbert, and G. Kathiresan, Eds., Boston: Kluwer Academic Publishers, pp. 283–313.

20. B. Murmann. 2013. Energy limits in A/D converters. *IEEE Faible Tension Faible Consommation (FTFC)*, Paris, France. pp. 1–4, June.

21. Y. Lim, K. Koh, K. Kim, H. Yang, J. Kim, Y. Jeong, S. Lee et al. 2010. A 1.1e- temporal noise 1/3.2-inch 8Mpixel CMOS image sensor using pseudo-multiple sampling. *International Solid-State Circuits Conference*, ISSCC, San Francisco. pp. 396–397, February.

22. M.-W. Seo, S.-H. Suh, T. Iida, T. Takasawa, K. Isobe, T. Watanabe, S. Itoh, K. Yasutomi, and S. Kawahito. 2012. A low-noise high intrascene dynamic range CMOS image sensor with a 13 to 19b variable-resolution column-parallel folding-integration/cyclic ADC. *IEEE Journal of Solid-State Circuits*, vol. 47, no. 1, pp. 270–283.

23. S. Matsuo, T. Bales, M. Shoda, S. Osawa, B. Almond, Y. Mo, J. Gleason, T. Chow, and I. Takayanagi. 2008. A very low column FPN and row temporal noise 8.9 M-pixel, 60 fps CMOS image sensor with 14bit column parallel SA-ADC. *Symposium on VLSI Circuits Digest of Technical Papers*, Honolulu, USA, June, pp. 138–139.

24. S. Yoshihara et al. 2006. A 1/1.8-inch 6.4 Mpixel 60 frames/s CMOS image sensor with seamless mode change. *IEEE Journal of Solid-State Circuits*, vol. 41, no. 12.

25. Takayuki et al. 2011. A 17.7Mpixel 120fps CMOS image sensor with 34.8 Gb/s readout. *International Solid-State Circuits Conference, ISSCC*.

26. J.-H. Kim et al. 2012. A 14b extended counting ADC implemented in a 24Mpixel APS-C CMOS image sensor. *International Solid-State Circuits Conference, ISSCC*, San Francisco. pp. 390–392, February.

27. M. Sasaki, M. Mase, S. Kawahito, and Y. Tadokoro. 2007. A wide-dynamic-range CMOS image sensor based on multiple short exposure-time readout with multiple-resolution column-parallel ADC. *IEEE Sensors Journal*, vol. 7, no. 1, pp. 151–158.

28. M. W. Seo et al. 2013. A low noise wide dynamic range CMOS image sensor with low-noise transistors and 17b column-parallel ADCs. *IEEE Sensors Journal*, vol. 13, no. 8, pp. 2922–2929, August.

29. M. S. Shin, J.-B. Kim, M.-K. Kim, Y.-R. Jo, and O.-K. Kwon. 2012. A 1.92 megapixel CMOS image sensor with column-parallel low-power and area-efficient SA-ADCs. *IEEE Transactions on Electron Devices*, vol. 59, no. 6, pp. 1693–1700, June.

30. D. G. Chen, F. Tang, and A. Bermak. 2013. A low-power pilot-DAC based column parallel 8b SAR ADC with forward error correction for CMOS image sensors. *IEEE Transactions on Circuits and Systems I*, vol. 60, no. 10, pp. 2572–2583, October.

31. S.-J. Tsai, Y.-C. Chen, C.-C. Hsieh, W.-H. Chang, H.-H. Tsai, and C.-F. Chiu. 2012. A column-parallel SAR ADC with linearity calibration for CMOS imagers. *IEEE Sensors Conference*, Taipei. pp. 1–4, October.

32. F. Tang, D. G. Chen, B. Wang, and A. Bermak. 2013. Low-power CMOS image sensor based on column-parallel single-slope/SAR quantization scheme. *IEEE Transactions on Electron Devices*, vol. 60, no. 8, pp. 2561–2566, August.

33. D. Chen, F. Tang, M.-K. Law, X. Zhong, and A. Bermak. 2014. A 64 fj/step 9-bit SAR ADC array with forward error correction and mixed-signal CDS for CMOS image sensors. *Circuits and Systems I: Regular Papers, IEEE Transactions on*, vol. 61, no. 11, pp. 3085–3093, November.

34. G. Emmert, E. Navratil, F. Parzefall, and P. Rydval. 1980. A versatile bipolar monolithic 6-bit A/D converter for 100 MHz sample frequency. *IEEE Journal of Solid-State Circuits*, vol. 15, pp. 1030–1032, December Available at: http://www.eetimes.com /document.asp?doc_id=1255778.

35. B. E. Jonsson and R. Sundblad. 2008. ADCs for sub-micron technologies. *EE Times Europe*, pp. 29–30, April.

36. J. Markus, P. Deval, V. Quiquempoix, J. Silva, and G. C. Temes. 2006. Incremental delta-sigma structures for DC measurement: An overview. *IEEE Custom Integrated Circuits Conference (CICC)*, San Jose, CA. pp. 41–48, September.

4 Low-Noise Detectors through Incremental Sigma–Delta ADCs

Adi Xhakoni and Georges Gielen

CONTENTS

4.1   Introduction 71
4.2   Low Noise Through Column-Level Multiple-Sampling ADCs 73
4.3   First-Order ISD ADC 75
4.4   Higher-Order ISD 77
4.5   Coefficient Scaling... 82
4.6   Building Blocks 83
      4.6.1   Integrator 83
      4.6.2   Quantizer 85
      4.6.3   DAC 85
4.7   PTC-Based ISD ADCs 85
4.8   Conclusions.. 88
References... 89

4.1   INTRODUCTION

Given their low cost, complementary metal–oxide semiconductor (CMOS) image sensors have had a rapid growth in recent years due to the massive use of cameras in consumer, industrial, and scientific applications such as mobile phones, machine vision, security, automotive, etc. The trend in imaging is the improvement of the image quality by increasing the spatial resolution of the sensor, by extending its dynamic range (DR), by increasing its frame rate, and by reducing the power consumption and the overall cost.

As the pixel array resolution increases, the simultaneous achievement of a high DR, a high frame rate, and low noise require highly parallel readout circuits [1]. The most common parallel readout architecture is based on column-level analog-to-digital converters (ADCs), as seen in Figure 4.1. Compared to conventional imagers with a single ADC per chip, the column-parallel architecture reduces the speed requirements of the ADCs proportionally to the number of columns. Furthermore, column-parallel readout architectures can achieve very low readout noise when the ADC is preceded by a low-pass filter and/or when multiple analog-to-digital conversions are performed per pixel, and the average value is taken.

M × N-pixels

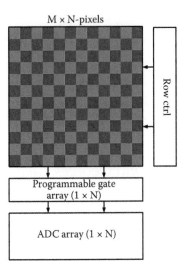

Programmable gate
array (1 × N)

ADC array (1 × N)

FIGURE 4.1 Block scheme of a traditional M × N-pixel image sensor with column-parallel readout circuitry.

The ADC characteristics are of crucial importance in the performance of the imager: the ADC consumes an important fraction of the total imager power [2] and can be the bottleneck in determining the frame rate, the noise, and the DR of the sensor.

A column-level readout with a sample-and-hold (S&H) block in front of the ADC is shown in Figure 4.2. This configuration allows a pipelined readout where, during the conversion of pixel n with its signal stored at the S&H, pixel $n + 1$ is being accessed and charges the column capacitance. As soon as pixel n is converted, the signal of pixel $n + 1$ is immediately stored at the S&H and can be digitized by the ADC. The ADC is always active, and in case of a digital-correlated double sampling (DCDS) readout, the pixel readout time is approximately twice the ADC conversion time. This is a case when the frame rate of the sensor is directly related to the ADC speed.

Recent pixel architectures based on pinned photodiodes can provide more than 90-dB inherent DR (for instance, 1-e⁻ noise and 40,000-e⁻ full well) [3]. In order to

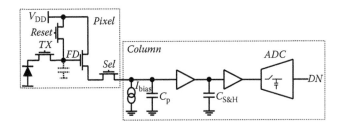

FIGURE 4.2 Schematic of a 4T pixel and its correspondent column-level readout. An S&H block is placed in front of the ADC to allow a pipelined readout of the reset level of the pixel and that of the signal level.

read this high DR signal, an ADC with a high resolution (e.g., higher than 14 bits) and a very-low-noise column readout circuit are needed. This poses great challenges as it is difficult to implement such a demanding column-level circuit within the ever-decreasing column pitches.

This results in two design trade-offs:

1. High-precision analog blocks are needed, which translates into a large ADC area. Therefore, more columns have to be shared by the same ADC, reducing the parallelism of the image sensor.
2. A longer conversion time is needed to achieve a high resolution: together with the decreased parallelism, this reduces the frame rate of the sensor.

An example of a CMOS imager with column-level readout based on high-resolution folding–cycling ADCs together with high-full-well pixels has been presented [4]. The reported performance included an 82-dB DR and a 1.2-e$^-$ noise at a low frame rate of 4 fps and at a relatively low spatial resolution of 1024 × 1032 pixels.

New column-level readout architectures are therefore necessary for the simultaneous achievement of a high spatial resolution, a high frame rate, and a high DR. This is translated into a large need in industry for column ADCs with high resolution, low noise, and low conversion time, and that consume a low area and a low power.

In this contest, incremental sigma–delta (ISD) ADCs are recently gaining interest in the imaging field as column-parallel converters [2,5]. This ADC architecture can achieve a high resolution (e.g., >14 bits) with low-precision and hence low-power and -area analog building blocks [6]. Furthermore, as ISD ADCs intrinsically use oversampling for the conversion, the thermal noise of the ADC and that of the pixel source follower (SF) are reduced as well [5]. The oversampled thermal noise allows the use of a smaller ADC sampling capacitance, occupying less area when compared to the Nyquist-type ADCs.

The low-precision analog building blocks needed allow the use of sigma–delta converters in high-speed imaging as well. A 14-bit resolution has been achieved at an impressive conversion time of 0.4 μs [2] with a third-order ISD ADC.

With this background, the aim of this chapter is to provide theoretical and practical insights in the design of column-level incremental sigma–delta ADCs for CMOS image sensors. This chapter is organized as follows. Section 4.2 deals with the effect of the oversampling on the pixel and on the column thermal noise. Section 4.3 introduces the first-order ISD ADC, while Section 4.4 deals with higher-order modulators. Design insights in the coefficient scaling and the analog building blocks of ISD ADCs are provided in Sections 4.5 and 4.6, respectively. Finally, a method to increase the conversion speed based on the photon transfer curve (PTC) is presented in Section 4.7.

4.2 LOW NOISE THROUGH COLUMN-LEVEL MULTIPLE-SAMPLING ADCs

Several noise sources contribute together to determine the minimum light level, which can be detected by a CMOS image sensor. Often, the dominant noise sources are the thermal noise of the pixel SF and that of the column readout circuitry [4]. We will analyze in this section the effect of oversampling on reducing these noise sources.

The readout path from a pixel to the column-level readout circuit, based on a multiple-sampling ADC, is shown in Figure 4.3. The capacitive load of the pixel SF is composed of the column parasitic capacitance C_p and of the sampling capacitor C_s of the multiple-sampling ADC.

DCDS is used for removing the pixel reset and fixed-pattern noise (FPN) and the vertical FPN of the column readout circuitry. The temporal noise at the output of the pixel SF in DCDS mode is expressed as [7]

$$\overline{n^2} \simeq \left(\frac{F_{SF}\xi_{SF}K_BT}{C_p+C_s} + \frac{K_BT}{C_s} + \overline{V_{n,ADC}^2} \right) \cdot \frac{2}{M} \tag{4.1}$$

where F_{SF} and ξ_{SF} indicate the closed-loop gain and the noise excess factor of the pixel SF [7], C_s and C_p represent the sampling capacitance of the first stage of the ADC and the parasitic capacitance at the output of the pixel SF, respectively, and M represents the number of samplings of the ADC (i.e., number of clock cycles). The first term of the equation represents the thermal noise contribution of the pixel SF, the second term is the thermal noise of the switch of the sampling capacitance of the ADC, and the last term is the thermal noise of the other blocks of the ADC. As expected, Equation 4.1 shows that the thermal noise reduces proportionally with the number of samplings M. When keeping the same noise level, the sampling capacitance C_s of the ADC can be reduced proportionally to M, when compared to a

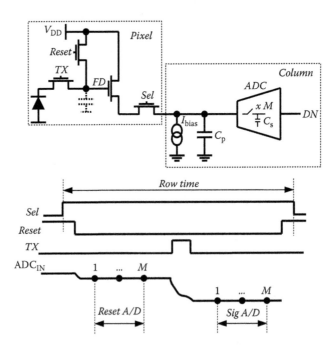

FIGURE 4.3   A multiple-sampling ADC is used for reducing the thermal noise of the pixel SF and that of the column circuitry.

conventional Nyquist ADC. Reducing this capacitance translates into a smaller ADC footprint. The major drawback of a multiple-sampling ADC is the increase in the conversion time due to the multiple analog-to-digital conversions that are needed.

The incremental sigma–delta ADCs intrinsically use oversampling for a single analog-to-digital conversion [8]. Therefore, these converters perform pixel and ADC thermal noise reduction without requiring multiple analog-to-digital conversions. This feature allows the combination of low noise with a high readout speed, i.e., with a high frame rate.

4.3 FIRST-ORDER ISD ADC

A first-order ISD ADC is shown in Figure 4.4. It is composed of a switched-capacitor integrator, a single-bit quantizer in the form of a comparator, and two reference voltages (V_{ref+} and V_{ref-}) representing a single-bit digital-to-analog converter (DAC). The coefficient $b1$ represents the gain of the integrator, which corresponds to the ratio between the sampling and the integrating capacitors of the integrator. The integrator $I(z)$ has the following transfer function:

$$I(z) = \frac{z^{-1}}{1 + z^{-1}} \tag{4.2}$$

The analog blocks of this ADC topology form the so-called modulator. The bit stream at the output of the modulator is processed by the digital filter. In the case of a first-order ISD ADC, the digital filter can be a simple ripple counter, which provides the digital representation (DN) of the analog input signal V_{in}. A circuit-level representation of the first-order ISD ADC is shown in Figure 4.5.

While similar to the conventional sigma–delta ADC, the incremental ADC converts signals that do not vary during the conversion. Such signals are commonly delivered by 4T pixels with pinned photodiodes and by S&H circuits. Furthermore, the incremental converter has a simpler digital filter than the conventional sigma–delta converter [9] and suffers less from instability as the modulator is reset before each conversion.

The first-order ISD ADC works as follows. The first operation consists of the reset of the integrator and that of the digital filter. The input signal V_{in} is then integrated first, and the output of the integrator is quantized by the comparator. According to

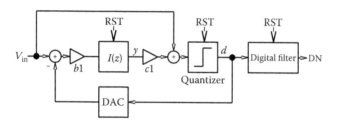

FIGURE 4.4 Block scheme of a first-order ISD ADC.

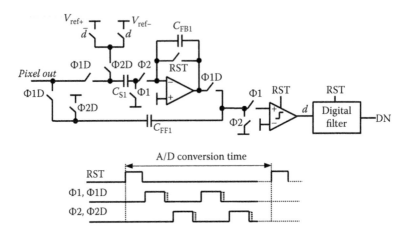

FIGURE 4.5 Circuit-level schematic and corresponding timing diagram of the first-order ISD ADC.

the quantization bit value d, V_{ref+} or V_{ref-} is subtracted from the input signal in the next clock phase.

This operation avoids the saturation of the integrator's output, keeping the modulator in a stable working condition. The output of the integrator is as follows:

$$y[0] = 0$$

$$y[1] = b1(V_{in}[0] - d_{[0]}[V_{ref+} - V_{ref-}])$$

$$y[M] = b1 \sum_{k=0}^{M-1} (V_{in}[k] - d_{[k]}[V_{ref+} - V_{ref-}]) \tag{4.3}$$

where M represents the clock cycles, i.e., the number of samplings, and d represents the output bit of the quantizer. According to Equation 4.3, the output of the integrator (y_{min} to y_{max}) is always within V_{ref-} and V_{ref+} if V_{in} is within the V_{ref-} and V_{ref+} voltage range as well. In order to simplify the expressions, from now on, we will use $V_{ref-} = 0$ and $V_{ref+} = V_{ref}$. With this assumption,

$$y_{min} < y[M] < y_{max}$$

$$0 < y[M] < V_{ref}$$

$$0 < b1 \cdot V_{in} \cdot M - b1 \sum_{k=0}^{M-1} (d_{[k]} \cdot V_{ref}) < V_{ref} \tag{4.4}$$

$$0 < V_{in} - \frac{1}{M} \sum_{k=0}^{M-1} (d_{[k]} \cdot V_{ref}) < \frac{V_{ref}}{b1 \cdot M}$$

This shows that the estimation of the input signal is equivalent to

$$\hat{V}_{in} = \frac{1}{M} \sum_{k=0}^{M-1} (d_{[k]} \cdot V_{ref})$$ (4.5)

It follows that the least significant bit (LSB) of the ADC is

$$LSB = \frac{V_{ref}}{b1 \cdot M}$$ (4.6)

The DN of the input signal can be achieved by the ripple counter at the output of the quantizer:

$$DN = \sum_{k=0}^{M-1} d_{[k]}$$ (4.7)

Requiring more than $\log_2 M$ clock cycles for N-bit resolution, the first-order ISD ADC is very slow for most of the imaging applications. For a 12-bit ideal resolution, 4097 clock cycles are needed when $b1 = 1$ is used. For instance, at a 50-MHz clock speed, the conversion time is approximately 82 μs, a high value for a column-level ADC. Given the gain–bandwidth requirements of the operational amplifier (op amp) of the integrator, the maximum clock speed is limited to a few tens of megahertz, preventing higher readout speeds. This conversion time is much higher than that of the single-slope ADCs, which can use clock speeds above 1 GHz [10]. For this reason, the first-order ISD ADC has had very limited application in commercial imagers except for a few pixel-level implementations.*

4.4 HIGHER-ORDER ISD

In order to decrease the number of clock cycles required per number of bits of resolution, higher-order ISD ADCs can be used. The higher-order modulator can be defined in z domain as follows [9,11]:

$$Y(z) = z^{-k} U(z) + (1 - z^{-1})^{La} E(z)$$ (4.8)

where $U(z)$ is the normalized input voltage, $E(z)$ is the quantization error of the comparator, La is the order of the modulator, and $k \le La$ is an integer.

Figure 4.6 shows a second-order ISD ADC with a feedback (FB) type of modulator [12]. As in the case of the first-order ISD ADC, the output of the last integrator can be used to determine the required number of clock cycles that are needed

---

* Insilixa dna testing sensor.

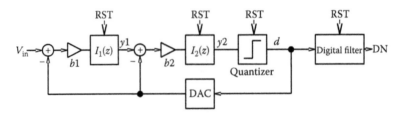

FIGURE 4.6   A second-order FB ISD ADC.

per conversion. Similarly to the first-order ISD ADC, the output of the first integrator is

$$y1[M] = b1\left(\sum_{k=0}^{M-1} V_{in}[k] - d_{[k]}V_{ref}\right) \tag{4.9}$$

while that of the second integrator is

$$y2[M] = b2\sum_{k=0}^{M-1} y1[k] - b2\sum_{k=0}^{M-1} d_{[k]}V_{ref}$$

$$= b1 \cdot b2\left(V_{in}\frac{M(M-1)}{2} - V_{ref}\sum_{i=0}^{M-1}\left(\sum_{k=0}^{i-1} d_{[k]} - \frac{d_{[i]}}{b1}\right)\right) \tag{4.10}$$

To ensure stability, the coefficients $b1$ and $b2$ are chosen such that the output of the second integrator never exceeds the voltage range V_{ref-} to V_{ref+} when the input voltage is within the V_{ref-} to V_{ref+} range. It follows that

$$0 < y2[M] < V_{ref} \tag{4.11}$$

Similarly to Equation 4.5, the estimation of the input is [13]

$$\hat{V}_{in} \approx \frac{2}{M(M-1)} V_{ref}\sum_{i=0}^{M-1}\left(\sum_{k=0}^{i-1} d_{[k]} - \frac{d_{[i]}}{b1}\right) \tag{4.12}$$

while the quantization error is

$$q_e = \frac{2 \cdot V_{ref}}{M(M-1) \cdot b1 \cdot b2} \tag{4.13}$$

As seen in Equation 4.12, the exact value of the modulator's coefficient $b1$ is needed to estimate the value of the analog input. Due to matching issues, large capacitors or column-specific calibration is needed. Furthermore, the output swing of both integrators can reach the limits of the V_{ref-} to V_{ref+} voltage range, requiring op amps with high voltage swing and increasing the LSB value, as seen in Equation 4.13.

To alleviate the drawbacks of the FB modulator, a feed-forward (FF) counterpart can be used [14]. A second-order FF ISD ADC is shown in Figure 4.7. In the FF architecture, the input signal is connected directly to the input of the quantizer.

In the time domain, the output of the second integrator after M clock cycles is as follows:

$$y2[M] = b2 \cdot b1 \sum_{i=0}^{M-1} \sum_{k=0}^{i-1} (V_{in[k]} - d_{[k]} V_{ref}) \tag{4.14}$$

Assuming again a constant input signal,

$$\sum_{i=0}^{M-1} \sum_{k=0}^{i-1} V_{in[k]} = \frac{M(M-1) \cdot V_{in}}{2} \tag{4.15}$$

As for stable operation, the output of the integrators is within the reference voltage range:

$$0 < y2[M] < V_{ref}$$

$$0 < V_{in} - \frac{2}{M(M-1)} \sum_{i=0}^{M-1} \sum_{k=0}^{i-1} d_{[k]} V_{ref} < \frac{2 \cdot V_{ref}}{M(M-1) \cdot b1 \cdot b2} \tag{4.16}$$

The estimation of the input voltage is

$$\hat{V}_{in} = \frac{2}{M(M-1)} \sum_{i=0}^{M-1} \sum_{k=0}^{i-1} d_{[k]} V_{ref} \tag{4.17}$$

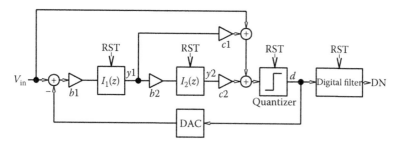

FIGURE 4.7 A second-order FF ISD ADC.

while the quantization error is now

$$q_e = \frac{2 \cdot V_{ref}}{M(M-1) \cdot b1 \cdot b2} \tag{4.18}$$

Note that we assumed a worst case scenario with the maximum last integrator output being V_{ref}. However, in a correctly designed FF modulator, the output of the last integrator is only a fraction of the V_{ref-} to V_{ref+} voltage range. It follows that the quantization error of second-order FF ISD ADCs is typically lower than the one that is expressed in Equation 4.18 and can be determined only through extensive behavioral and circuit simulations.

As in the case of the first-order modulator, the input voltage estimation does not include the coefficient values of the modulator. This is a great advantage compared to the FB architecture as the exact value of the capacitors is not needed to estimate the input voltage. Furthermore, as the input is not processed by the low-pass filter but fed to the input of the quantizer, the voltage swing of the integrators is limited to a fraction of the full-swing V_{ref-} to V_{ref+} voltage range. This corresponds to a simpler integrator design with the potential for a lower power consumption and a lower quantization noise, as from Equation 4.18. Due to the characteristics above, we will only discuss FF ISD ADC architectures in this chapter.

A digital filter for a second-order ISD ADC has been proposed [5]. Further explanations are present in the patent application [15].

Assuming, ideally, unity gain integrators and full ADC input voltage swing (V_{ref-} to V_{ref+}), Equation 4.18 shows that in order to reach a 12-bit ADC resolution, less than 100 clock cycles are now needed. This is a substantial improvement compared to the more than 4097 clock cycles that are required by the first-order ISD ADC.

An even lower number of clock cycles can be achieved by increasing the order of the modulator and that of the digital filter. Figure 4.8 shows a third-order ISD ADC.

Proceeding with the same calculations as in Equation 4.14, the output of the third integrator is

$$y3[M] = b3 \cdot b2 \cdot b1 \sum_{j=0}^{M-1} \sum_{i=0}^{j-1} \sum_{k=0}^{i-1} \left(V_{in[k]} - d_{[k]} V_{ref} \right) \tag{4.19}$$

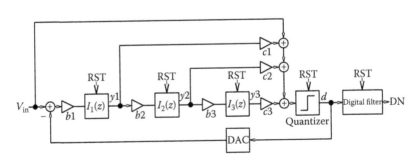

FIGURE 4.8   A third-order FF ISD ADC.

with

$$\sum_{j=0}^{M-1}\sum_{i=0}^{j-1}\sum_{k=0}^{i-1}V_{\text{in}[k]} = \frac{M(M-1)(M-2)\cdot V_{\text{in}}}{3!} \tag{4.20}$$

Limiting $y3[M]$ between 0 and V_{ref} gives the following estimation of the input voltage $\hat{V}_{\text{in}}$:

$$\hat{V}_{\text{in}} = \frac{3!\cdot V_{\text{ref}}}{M(M-1)(M-2)}\sum_{j=0}^{M-1}\sum_{i=0}^{j-1}\sum_{k=0}^{i-1}V_{\text{in}[k]} \tag{4.21}$$

The quantization error value is

$$q_{\text{e}} = \frac{3!\cdot V_{\text{ref}}}{M(M-1)(M-2)\cdot b3\cdot b2\cdot b1} \tag{4.22}$$

While the maximum output code is greatly increased compared to that of the second-order ISD ADC, the quantization error contains a third coefficient $b3$, reducing its efficacy a bit. As stability is a greater concern for higher-order modulators, conservative coefficient scaling is now required, and the input voltage has to be scaled with respect to the reference voltages. More details on the stability and the coefficient choice are explained in Section 4.5.

Increasing the modulator order reduces the required number of clock cycles that are needed for a certain resolution. However, the price paid is the increase in area and especially the reduction of the averaging of the noise.

The digital filter of the ADC has a cycle-dependent weight coefficient [5]. In a first-order ISD ADC, after each sampling of the input signal, the output of the integrator is quantized, and the quantization bit is counted by a ripple counter. As for M sampling, there are M quantization levels; each quantization bit and each sampling have the same impact or weight on the total digital count, and Equation 4.1 is still valid. For higher-order ISD ADCs, each sampling has a different impact on the total digital number at the output. For instance, for a second-order ISD ADC with M clock cycles, the first quantization bit of the quantizer is counted M times, the second bit is counted $M-1$ times, and so on. Equation 4.1 is modified as follows:

$$\overline{n^2} \simeq \left(\frac{F_{\text{SF}}\xi_{\text{SF}}K_{\text{B}}T}{C_{\text{p}}+C_{\text{s}}} + \frac{K_{\text{B}}T}{C_{\text{s}}} + \overline{V_n^2}_{\text{,ADC}}\right)\frac{2}{M}\cdot n_{\text{rd}} \tag{4.23}$$

where n_{rd} represents the noise reduction factor. An n_{rd} of 4/3 for the second-order ISD has been calculated by [5], and an n_{rd} of 1.8 for a third-order ISD has been calculated by [9]. It follows that the lower the order, the better the noise averaging of the ISD

ADC. A lower sampling capacitor can be used in lower-order modulators due to the calculation in Equation 4.23. Given its good compromise between area, speed, noise reduction, and stability, the second-order ISD ADC seems to be best suited for column-level implementation. In Section 4.7, a method will be presented to further reduce the number of clock cycles for a required resolution.

4.5   COEFFICIENT SCALING

In order to avoid instability, i.e., avoid the output of the integrators to saturate, the coefficients of the modulators have to be scaled, and the input voltage has to be reduced to a fraction of the DAC reference voltage range. However, scaling the coefficients increases the quantization error (see Equation 4.18), while reducing the input voltage range reduces the DR. A trade-off is present between the two methods.

In imaging applications, the area of the modulator is of crucial importance, and the pitch of the ADC has to be equal to that of the pixel. Nevertheless, the ADC area is constrained by the magnitude of the coefficients for the following reasons. The coefficients are determined by the ratio between the sampling capacitor and the FB capacitor of the integrator. While the sampling capacitor is fixed by the noise requirement (see Equation 4.23), the FB capacitor is equivalent to C_s/b where b is the coefficient of the integrator, and C_s is the sampling capacitor. It follows that a small coefficient increases the size of the FB capacitor and therefore the total area of the ADC. Extensive behavioral simulations are needed to find the optimal coefficient and input voltage range. In imagers where the ADC area is a major constraint (e.g., in imagers with small pixel pitch), reducing the input voltage swing is the preferred option. In such cases, the coefficients are determined by the area constraint, while the input voltage is increased by small steps to determine the maximum value before instability.

The FF coefficients c_i are chosen in order to satisfy Equation 4.8. When b_i coefficients scale for stability reasons, the c_i coefficients should scale as well. If $b_i' < 1$, then $c_i' = c_i/b'$.

As opposed to the FB modulators, FF modulators present a summing point where the FF branches are summed together. Figure 4.9 shows the implantation of the switched capacitor summing circuit. C_{FFi} represents the capacitance of the coefficient C_i. The transfer function of this circuit in z domain can be expressed as

$$Y(z) = \frac{\displaystyle\sum_{i=1}^{n} X_i(z) \cdot C_{FFi}}{\displaystyle\sum_{i=1}^{n} C_{FFi}} \tag{4.24}$$

As seen in Equation 4.24, FF coefficients $c_i > 1$ are not possible with this circuit but require active amplifiers. However, at the input of the single-bit quantizer, only the sign of the signal is needed, whereas the information about the amplitude can be

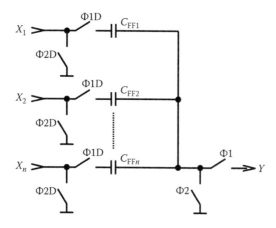

FIGURE 4.9 Switched-capacitor–based summing circuit used for implementing the FF coefficients C_n.

omitted. It follows that as the denominator of Equation 4.24 represents an amplification, it can be omitted as well, and Equation 4.24 becomes

$$Y(z) = \sum_{i=1}^{n} X_i(z) \cdot C_{\text{FF}i} \tag{4.25}$$

4.6 BUILDING BLOCKS

The integrators, the quantizer, and the DAC are analyzed in detail in this section.

4.6.1 INTEGRATOR

In the switched-capacitor implementation of the ISD ADC, the integrators are implemented using a stray-insensitive architecture [16]. Nonoverlapping clocks are used to control the sampling and the integration phase. A switch connects the output and the negative input of the op amp to perform the reset of the integrator at the beginning of the conversion process.

The modulators are usually designed to be thermal noise limited [11]. The thermal noise is typically set by the sampling capacitor of the first integrator: due to oversampling, relatively small values of sampling capacitor can be used (e.g., 50 fF), reducing the load of the op amp and therefore reducing its power consumption. Small capacitor values are possible in FF architectures since, as seen in Equations 4.17 and 4.20, the estimation of the input signal does not depend on the value of the coefficients.

Together with the DAC reference voltages, the op amp in the integrator is the most power-consuming building block of the entire ISD ADC [12]. The important specifications of the op amps are the DC gain, the gain bandwidth, and the noise. Extensive behavioral simulations are needed to evaluate the impact of these characteristics on the performance of the converter.

The design of the op amp of the first integrator is critical. The successive integrators have a lower impact on the performance of the modulator due to the noise-shaping characteristics of the modulator [11]. The common practice is a factor of 4 in power reduction and noise increase of the op amps of the successive integrators.

The finite gain of the op amps introduces leakage in the integrators and is an important error source [8]. The minimum detectable signal by a first-order modulator and by a second-order modulator is $1/2A$ and $1.5/A^2$, respectively, where A represents the gain of the op amp [11]. It follows that higher-order ISD ADCs relax the requirements on analog building blocks.

The op amp should be designed to limit the thermal noise contribution, i.e., its noise should be lower than the sampling noise [17]. Typically, single-stage op-amp architectures are chosen due to their good trade-off in power, speed, and noise performance [17]. To achieve sufficient gain, single-stage cascode op-amp topologies are typically used [18]. Given their low output voltage swing, these topologies are well compatible with FF modulators.

The gain–bandwidth of the op amp of the first integrator is typically set as five times the sampling frequency. Again, due to the noise-shaping feature of the modulators, the op amps in the remaining integrators of the modulator can be made slower (approximately four times), reducing the power consumption.

The common-source cascode amplifier (Figure 4.10) is a conventional architecture that is used in previous scientific imager works [19]. Given the single current branch, low power and low area can be achieved. However, it suffers from a reduced power supply rejection ratio (PSRR), demanding a very clean supply voltage that is difficult to achieve at column level.

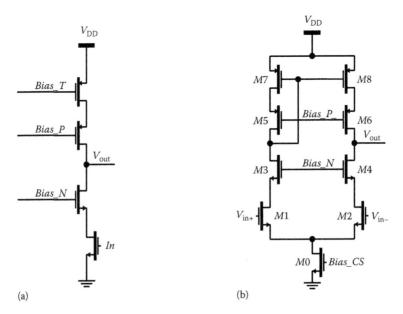

(a)   (b)

FIGURE 4.10   Common source (a) and telescopic cascode (b) op amp architectures.

The telescopic cascode architecture (Figure 4.10) provides a high gain due to the increased output impedance by cascoding more transistors at the output node. The decreased output voltage swing, one of the main drawbacks of this architecture, is not a problem due to the use of an FF modulator architecture. Furthermore, it has a good PSRR due to the differential input, making it an ideal candidate for column-level ISD ADCs.

4.6.2 QUANTIZER

The quantizer is implemented through a dynamic comparator [12]. To reduce the kickback effect of the comparator to the previous stages of the ADC, a preamplifier is used. The preamplifier is also useful in reducing the offset of the comparator. The offset of the comparator is mitigated by the use of the DCDS technique, which is intrinsically utilized for converting the reset and the signal level of the pixel.

4.6.3 DAC

The DAC of the modulator of the column ISD ADC is composed of two reference voltages, which are shared between multiple-column ADCs. This allows a high gain uniformity between columns without the need for column-specific gain calibration. The drawback is that the reference voltages drive the sampling capacitors of multiple ADCs, requiring power-consuming voltage buffers and wide reference distribution metal lines for low-noise operation. To reduce the power of the reference buffers, column-specific buffers based on SFs have been suggested in a previous work [20].

4.7 PTC-BASED ISD ADCs

We now address the excessive number of clock cycles per conversion that are needed by low-order ISD ADCs. The problem can be tackled by exploiting one of the main issues of the photodiodes: the photon shot noise.

The photodiodes are subject to signal-dependent photon shot noise. The photon shot noise, expressed as number of electrons, is $\sqrt{Q}$ with Q being the number of charges that are generated by the photodiode. Modern pixels provide a maximum signal/noise ratio (SNR) of approximately 40–45 dB (Figure 4.11). However, they can provide more than 90-dB DR when in the presence of a large full well and low noise at low light.

As seen in Figure 4.11, a high ADC resolution is only needed at low light. At high-light levels, a lower ADC resolution can be used, with its benefits being the power consumption and the conversion time.

Common ADCs with SNR ≈ DR are not optimal in image sensors as they waste power and conversion time for the extra SNR. Image sensor ADCs applying a conversion method based on the PTC such as single slope [21], SAR [22], etc., have shown important improvements in terms of speed and/or power consumption compared to their SNR ≈ DR counterpart. For instance, [21] shows a three-time improvement in the speed of the PTC-based single-slope ADC compared to the conventional single-slope ADC, without an increase in power consumption.

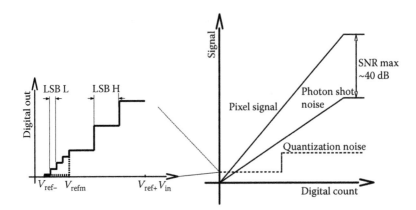

FIGURE 4.11 The photon shot noise dominates at high light; therefore, a small ADC quantization step (LSB L) is required only at low light.

An ISD ADC with second-order FF modulator implementing a PTC-based conversion has been proposed by [18] and works as follows. Ideally, the ADC resolution N in bits provided by the second-order digital filter is

$$N = \log_2 (M(M + 1)) - 1 \qquad (4.26)$$

where M is the number of clock cycles. According to Equation 4.26 and considering that the input range of the ISD ADC depends on the DAC levels, the quantization step Q_s can be expressed as

$$Q_s = \frac{2(V_{ref+} - V_{ref-})}{M(M + 1)} \qquad (4.27)$$

where V_{ref+} and V_{ref-} represent the high and low DAC reference voltages, respectively.

Figure 4.12 shows the simplified schematic of the column circuitry including the PTC-based ISD ADC. The main differences compared to the conventional ISD ADC are the extra comparator at the input of the ADC and the addition of a third reference voltage named V_{refm} having a voltage level in between V_{ref-} and V_{ref+}. Before each conversion, the input comparator detects if the ADC input signal is below a certain threshold voltage. V_{refm} or V_{ref+} is chosen at low or high light, respectively, depending on the comparison. Having a V_{refm} with a value close to V_{ref-} allows a low quantization step at low light, as seen in Equation 4.27.

The equivalent ADC DR in bits is increased by

$$DR_{EXT} = \log_2 \left(\frac{V_{ref+} - V_{ref-}}{2(V_{ref+} - V_{ref-})} \right) \qquad (4.28)$$

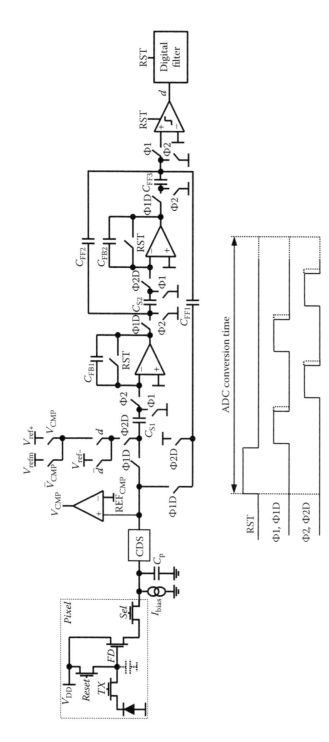

FIGURE 4.12 PTC-based ISD ADC schematic and timing diagram. The modulator is an FF second-order ISD ADC. The PTC conversion is achieved by adding a comparator and an extra DAC level, V_{refm}.

A typical configuration with V_{ref-}, V_{refm}, and V_{ref+}, respectively, 1, 1.2, and 2.6 V corresponds to 3-bit DR extension at low light. According to Equations 4.26 and 4.28, 12-bit equivalent ADC resolution can be achieved with only 35 clock cycles, drastically reducing the 100 clocks that are needed by the conventional second-order ISD ADC. Due to coefficient scaling, the number of clock cycles is generally increased above 40. At high light, the resolution is 9 bits. As at high light, the SNR of the pixel is limited to approximately 40 dB by the photon shot noise, a 9-bit ADC is sufficient. Therefore, in imaging applications, this PTC ISD ADC provides nearly the same performance as a conventional 12-bit ISD ADC, be it only at a fraction of the number of clock cycles.

The output of the added comparator is taken as the exponent bit, determining the chosen reference. As only 10 bits (9-bit ADC + 1-bit comparator) are used for representing the 12-bit ADC resolution, the method is also useful in reducing the digital data rate and simplifying the design of the digital filter of the ADC.

At the digital signal processor (DSP) or field-programmable gate array level, to restore the 12-bit coding of the pixel, the following method is used. If the comparator bit indicates that V_{refm} was used, the LSB of the digital output of the ADC is mapped onto the LSB of a 12-bit register. If V_{ref+} was used, then the most significant bit (MSB) of the digital output of the ADC is mapped onto the MSB of the 12-bit register. As the three reference voltages are shared by all the columns of the imager, no column-specific calibration is needed to avoid column FPN. The PTC method is therefore very effective in reducing the number of clock cycles of the ISD ADC, allowing a high frame rate and low power consumption.

4.8 CONCLUSIONS

In this chapter, we provided theoretical and practical insights on the incremental sigma–delta ADCs, an emerging class of converters for imaging applications. Different ISD ADC architectures were analyzed, and their trade-offs were discussed.

While easier to design and consuming the lowest area, the first-order ISD ADCs are slow for most of the imaging applications. Therefore, higher-order converters are advised as column-level ADCs. Furthermore, we saw that the FF architectures are superior to their feedback counterparts due to two reasons. First, the FF architectures have a lower voltage swing at the output of the integrator, allowing the use of low-power cascode op amps. Second, in FF architectures, the estimation of the input voltage does not require the precise knowledge of the scaling coefficients; therefore, no column calibration is needed.

A method to reduce the number of clock cycles per conversion was provided as well. The method is based on the PTC. The ISD ADC uses a third reference voltage and an extra comparator to modify the quantization step according to the input voltage level. This way, the number of clock cycles required for a given DR could be reduced by up to three times when compared to the classic ISD architecture with the same order.

REFERENCES

1. A. Xhakoni, H. Le-Thai, and G. Gielen. 2014. A low-noise high-frame-rate 1-d decoding readout architecture for stacked image sensors. *Sensors Journal, IEEE,* vol. 14, no. 6, pp. 1966–1973, June.

2. B. Cremers, M. Innocent, C. Luypaert, J. Compiet, I. C. Mudegowdar, C. Esquenet, G. Chapinal et al. 2013. A 5 megapixel, 1000fps CMOS image sensor with high dynamic range and 14-bit a/d converters. In *IISW,* Snowbird, USA, June 12–16, 2013.

3. B. Fowler, C. Liu, S. Mims, J. Balicki, W. Li, H. Do, J. Appelbaum, and P. Vu. 2010. A 5.5mpixel 100 frames/sec wide dynamic range low noise CMOS image sensor for scientific applications. *Proc. SPIE 7536, Sensors, Cameras, and Systems for Industrial/ Scientific Applications XI,* pp. 753 607–753 607–12. [Online]. Available at http://dx.doi .org/10.1117/12.846975.

4. M.-W. Seo, S.-H. Suh, T. Iida, T. Takasawa, K. Isobe, T. Watanabe, S. Itoh, K. Yasutomi, and S. Kawahito. 2012. A low-noise high intrascene dynamic range CMOS image sensor with a 13 to 19b variable-resolution column-parallel folding-integration/cyclic ADC. *Solid-State Circuits, IEEE Journal of,* vol. 47, no. 1, pp. 272–283.

5. Y. Chae, J. Cheon, S. Lim, M. Kwon, K. Yoo, W. Jung, D.-H. Lee, S. Ham, and G. Han. 2011. A 2.1 m pixels, 120 frame/s CMOS image sensor with column-parallel delta sigma ADC architecture. *Solid-State Circuits, IEEE Journal of,* vol. 46, no. 1, pp. 236–247, January.

6. J. Nakamura, B. Pain, T. Nomoto, T. Nakamura, and E. Fossum. 1997. On-focal-plane signal processing for current-mode active pixel sensors. *Electron Devices, IEEE Transactions on,* vol. 44, no. 10, pp. 1747–1758, October.

7. S. Kawahito. 2007. Signal processing architectures for low-noise high-resolution CMOS image sensors. In *Custom Integrated Circuits Conference, 2007, CICC '07, IEEE,* San Jose, USA, pp. 695–702, September 16–19, 2007.

8. J. Markus, J. Silva, and G. Temes. 2004. Theory and applications of incremental delta; sigma; converters. *Circuits and Systems I: Regular Papers, IEEE Transactions on,* vol. 51, no. 4, pp. 678–690, April.

9. J. Markus. 2005. *Higher-Order Incremental Delta-Sigma Analog-to-Digital Converters.* PhD Thesis, University of Budapest, Hungary.

10. I. Takayanagi and J. Nakamura. 2013. High-resolution CMOS video image sensors. *Proceedings of the IEEE,* vol. 101, no. 1, pp. 61–73.

11. R. Schreier and G. Temes. 2004. *Understanding Delta-Sigma Data Converters.* Wiley. [Online]. Available at http://books.google.be/books?id=5HNqQgAACAAJ.

12. L. Yao, M. S. J. Steyaert, and W. Sansen. 2004. A 1-v 140-mu;w 88-db audio sigma-delta modulator in 90-nm CMOS. *Solid-State Circuits, IEEE Journal of,* vol. 39, no. 11, pp. 1809–1818.

13. J.-E. Park, D.-H. Lim, and D.-K. Jeong. 2014. A reconfigurable 40-to-67 db SNR, 50-to-6400 Hz frame-rate, column-parallel readout IC for capacitive touch-screen panels. *Solid-State Circuits, IEEE Journal of,* vol. 49, no. 10, pp. 2305–2318, October.

14. R. Schreier. 1993. An empirical study of high-order single-bit delta-sigma modulators. *Circuits and Systems II: Analog and Digital Signal Processing, IEEE Transactions on,* vol. 40, no. 8, pp. 461–466, August.

15. Y. Chae. 2012. Decimation filters, analog-to-digital converters including the same, and image sensors including the converters. *Patent grant US 8233068 B2.*

16. J. Baker. 2011. *CMOS: Circuit Design, Layout, and Simulation.* John Wiley & Sons.

17. S. Porrazzo, F. Cannillo, C. Van Hoof, E. Cantatore, and A. van Roermund. 2012. A power-optimal design methodology for high-resolution low-bandwidth sc $5\ a$ modulators. *Instrumentation and Measurement, IEEE Transactions on,* vol. 61, no. 11, pp. 2896–2904, November.

18. A. Xhakoni, H. Le-Thai, T. Geurts, G. Chapinal, and G. Gielen. 2014. PTC-based sigma-delta ADCs for high-speed, low-noise imagers. *Sensors Journal, IEEE,* vol. 14, no. 9, pp. 2932–2933, September.

19. J. Aoki, Y. Takemoto, K. Kobayashi, N. Sakaguchi, M. Tsukimura, N. Takazawa, H. Kato et al. 2013. A rolling-shutter distortion-free 3D stacked image sensor with -160db parasitic light sensitivity in-pixel storage node. In *Solid-State Circuits Conference Digest of Technical Papers (ISSCC), 2013 IEEE International,* San Francisco, pp. 482–483, February 17–21, 2013.

20. A. Xhakoni and G. Gielen. 2013. Designing incremental sigma-delta ADCs for low thermal noise in image sensors. In *International Image Sensors Workshop,* Snowbird, USA, pp. 482–483, June 12–16, 2013.

21. A. Theuwissen. 2007. CMOS image sensors: State-of-the-art and future perspectives. In *Solid State Circuits Conference, 2007, ESSCIRC 2007, 33rd European,* September, pp. 21–27.

22. S. Huang, R. Tantawy, S. Lam, and D. V. Blerkom. 2011. Design of a PTC-inspired segmented ADC for high-speed column-parallel CMOS image sensor. In *IISW,* Hokkaido, Japan, June 8–11, 2011.

5 Time-to-Digital Conversion

Yasuo Arai

CONTENTS

5.1 Introduction 92
5.2 Basic TDC Performance Figures...93
    5.2.1 TDC Operation Mode..93
    5.2.2 Measurement Error 95
5.3 Various TDC Circuits 97
    5.3.1 Counter and Shift Register 98
    5.3.2 Time-to-Amplitude Converter 98
    5.3.3 Delay-Line TDC 99
    5.3.4 Pulse-Shrinking TDC.. 101
    5.3.5 Subgate Delay Resolution 102
        5.3.5.1 Phase Interpolator/Blender 104
        5.3.5.2 Vernier TDC 104
        5.3.5.3 Complex RO.. 105
    5.3.6 Implementation in an FPGA.. 105
    5.3.7 Custom TDC Chips 107
5.4 TDC Applications.. 109
    5.4.1 Column-Parallel ADC in Image Sensor 109
    5.4.2 ADC/TDC for Automobile 110
    5.4.3 Pixel TDC 110
        5.4.3.1 Single-Photon Avalanche Diode Detector 110
        5.4.3.2 Medipix and Timepix Detectors 111
    5.4.4 All-Digital Phase-Locked Loop 112
5.5 Advanced TDC Circuits 114
    5.5.1 Time Difference Amplifier 114
    5.5.2 Stochastic TDC.. 116
5.6 Summary 117
Acknowledgments.. 117
References... 117

5.1 INTRODUCTION

The analog-to-digital converter (ADC) is one of the most popular devices in electronics, but the time-to-digital converter (TDC) is a not-so-popular device. The TDC has been used mostly in scientific experiments to measure the lifetime of nuclei, the quenching time of fluorescence, the drift time of ions, the time of flight (ToF) of moving particles, and so on. In the meantime, the application of the TDC is becoming wider and wider, and many people are willing to know more about the TDC. Unfortunately, there are only a handful of references: (a) some commercial TDC devices, (b) review papers [1,2], and (c) a book [3] about TDC. This is mainly because the TDC circuit is mostly custom implemented in application-specific integrated circuits (ASICs) or field-programmable gate array (FPGA).

An example of the early implementation of a time difference measurement circuit is shown in Figure 5.1. This circuit was presented in the *Journal of the Physical Society Japan* in 1958 [4]. At that time, the only element that could produce nanosecond timing precisely was a cable. The circuit detects the collision point of two pulses within two long coaxial cables. The collision point indicates the time difference of two pulses, and it is converted into a delayed pulse and observed in an oscilloscope.

At present, the delay element is mostly implemented in the inverters or buffers of semiconductor devices. However, the basic idea is still surprisingly very similar to the above circuit.

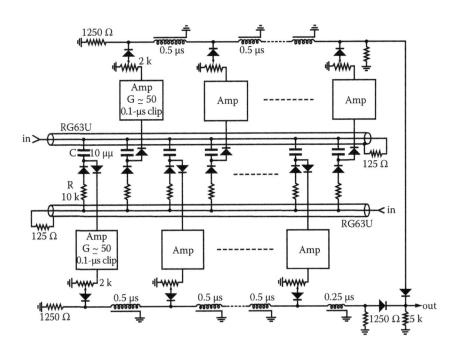

FIGURE 5.1 Time difference measuring circuit (Chronotron), which uses 12-m coaxial cables, developed in 1958. (Adapted from G. Tanahashi. 1958. *BUTSURI*, Vol. 13, No. 2, pp. 72–74. With Permission.)

Although there are very few commercial products of TDC [5,6], the time difference measuring circuit is becoming more and more important. One of the main applications of the TDC is the distance detection system of automobile and airplanes. Another example is time-of-flight positron emission tomography (ToF-PET).

In addition to the direct time measurement application, the TDC circuit is used to replace conventional analog circuitry. Since the operating voltage of large-scale integrated-circuits (LSIs) goes very low and the device speed becomes very high, the time domain operation becomes easier than traditional amplitude domain operation in analog circuit. For example, in mobile phones, the phase-locked loop (PLL) circuit was one of the remaining analog circuits where the phase detection of oscillator and reference clock is necessary. However, nowadays, the TDC circuit can replace the phase detection circuit, and the all-digital PLL circuit becomes popular and widely used in mobile devices.

5.2   BASIC TDC PERFORMANCE FIGURES

ADC and TDC look very similar, but they have many distinct differences. Figure 5.2 shows their basic digitization scheme. The ADC digitizes input signal amplitude with periodical sampling. Thus, the output data rate is normally constant. On the other hand, the TDC digitizes input signal transition time (leading edge or trailing edge or both), which is asynchronous to the digitizing system clock. Thus, we do not know when and how many output data will be generated. Thus, the volume of data depends on the input transition rate.

5.2.1   TDC OPERATION MODE

The TDC circuit can be classified into two categories depending on the calibration method: (1) continuous calibration (online) and (2) periodical calibration (offline) (Figure 5.3). A continuous calibration circuit refers to an external reference clock and continuously adjusts the delay time of delay elements to a fixed value in paralel with time measurements. Although the circuit becomes a bit more complex, this scheme is good for high-rate measurements that do not allow any dead time for calibration or for applications that require absolute time accuracy immediately.

On the other hand, the periodical calibration scheme is very simple. It measures a reference clock period occasionally. This assumes that the delay time of the delay

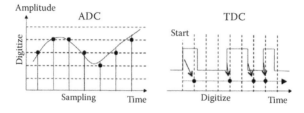

FIGURE 5.2   Basic operation of ADC and TDC. The ADC output has a constant rate, whereas the TDC output is not constant.

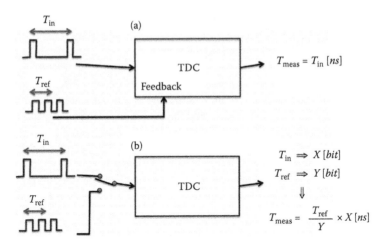

FIGURE 5.3 (a) Continuous calibration: TDC is continuously calibrated by referencing external clock period (T_{ref}). (b) Periodical calibration: The reference clock period (T_{ref}) is sampled periodically, and measured time is converted to a real value by using the reference clock period at a later stage.

elements does not change suddenly. Measured value is calibrated by using the data of a reference clock at a later stage. This scheme does not guarantee the minimum resolution of the circuit, but it is enough for slow sampling rate applications such as distance measurement. This enables an easy design and a cost-effective circuit.

The specification of a TDC can be characterized in several ways, as shown in Figure 5.4. Since the origin of the time must be defined someway, there are two possible schemes: (1) common start and (2) common stop measurement. In common start mode, the measurement is started at $t = 0$, and consecutive pulse times relative

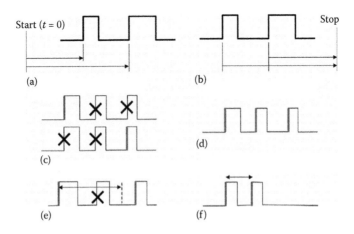

FIGURE 5.4 TDC characterization. (a) Common start and (b) common stop measurement. (c) Single-hit and (d) multihit TDC. (e) Conversion time and (f) double-pulse timing resolution.

to $t = 0$ are measured. In common stop mode, the circuit is started in advance, and the time from signal transitions to stop is measured. This is the typical mode in radiation measurements where the trigger signal will come later than the actual particle passing signal.

The TDC that can measure multiple times is called a multihit TDC, whereas the TDC that can measure only a single point is called a single-hit TDC (Figure 5.4c and d). Every TDC needs some conversion time (or dead time) (Figure 5.4e). Especially, the minimum time to separate two pulses is called double-pulse timing resolution (Figure 5.4f).

5.2.2 MEASUREMENT ERROR

If the start or stop time is synchronous to the system clock, the measurement error mainly comes from a single-hit measurement. Therefore, the measurement error of TDC is similar to that of an ADC in this case. Figure 5.5a shows the quantization error for an ideal TDC. The quantization error is distributed from −1/2 to +1/2 LSB uniformly; thus, the average quantization error $\Delta t_{\text{quantize}}$ is

$$\left\langle \Delta t^2_{\text{quantize}} \right\rangle^{0.5} = \left(\int_{-1/2}^{1/2} t^2 \, dt \right)^{0.5} = \frac{1}{\sqrt{12}} \approx 0.289 \, [\text{LSB}]$$

Differential nonlinearity (DNL) and integral nonlinearity (INL) are defined, as shown in Figure 5.5b. In a TDC, usually, the offset error is not important since the origin of $t = 0$ is determined arbitrary, and the gain (slope) error can be normally neglected if the right calibration method is used.

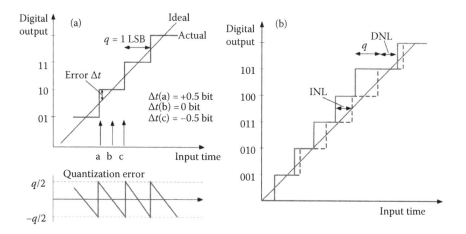

FIGURE 5.5   TDC conversion errors: (a) quantization error and (b) DNL and INL.

In the actual measurement, the additional measurement error will be added to the quantization error. If the measurement error has a Gaussian distribution of σ, the actual error distribution $F(\Delta t, \sigma)$ will be

$$F(\Delta t, \sigma) = \int_{-\infty}^{\infty} G(\Delta t, \mu, \sigma) P(\mu)\, d\mu = -\frac{1}{2q}\left(\operatorname{erf}\left(\frac{\Delta t - \frac{q}{2}}{\sigma\sqrt{2}} \right) - \operatorname{erf}\left(\frac{\Delta t + \frac{q}{2}}{\sigma\sqrt{2}} \right) \right)$$

$$\text{where } P(\mu) = \begin{cases} \frac{1}{q} & |\mu| \le \frac{q}{2} \\ 0 & |\mu| > \frac{q}{2} \end{cases}, \quad G(\Delta t, \mu, \sigma) = \frac{1}{\sigma\sqrt{2\pi}} \exp\left(-\frac{1}{2}\left(\frac{\Delta t - \mu}{\sigma} \right)^2 \right),$$

$$\operatorname{erf}(z) = \frac{2}{\sqrt{\pi}} \int_0^z e^{-t^2}\, dt \text{ (Error function)}$$

where q is the bit width. An example of time measurement error distribution is shown in Figure 5.6.

In many cases, we measure the time difference of pulses. If the measurements are done in two independent TDCs, the combined error follows the standard error formula; the error distribution becomes a triangular shape, and the root mean square (RMS) error becomes $1/\sqrt{6}$ (≈ 0.408) LSB. However, in most cases, the time difference is measured in a single TDC that uses a single clock. Therefore, the two measurements are not independent anymore.

As shown in Figure 5.7a, if the input time difference ($T_{in} = Q + F$, $Q = 0, 1, 2,...$, $0 < F < 1$) coincides with the TDC quantization value Q, and has no fractional value of the LSB ($F = 0$), the error is always 0 irrelevant to edge position. If the fractional value exists, the obtained value T_m will have the probability to be Q or $Q + 1$

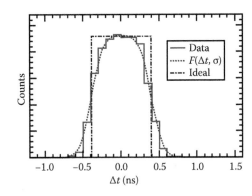

FIGURE 5.6 An example of error distribution for ideal Gaussian error ($F(\Delta t, \sigma)$) cases and actual measurement data (1 LSB = 0.78 ns).

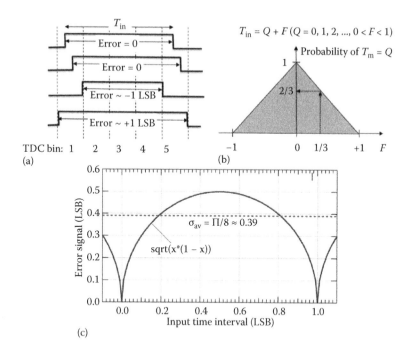

FIGURE 5.7 (a) Time interval measurement and error. (b) Probability distribution of $T_m = Q$ versus fractional time F. (c) Standard deviation versus input time interval.

depending on the value F. For example, if $F = 1/3$, two-thirds of the obtained value will be Q, and the another third will be $Q + 1$ (Figure 5.7b). Thus, the standard deviation σ changes depending on T_{in}, as shown in Figure 5.7c, and the average standard deviation will be $\pi/8 \approx 0.393$ LSB.

If we repeat the measurement N times for the same input signal, and get the results T_1 and T_2 with the numbers N_1 and N_2, respectively ($N_1 + N_2 = N$), the averaged value T_N and the average standard deviation $\sigma_{N\,av}$ are

$$T_N = \frac{T_1 N_1 + T_2 N_2}{N}, \sigma_{N\,av} = \frac{0.39}{\sqrt{N}} \, [\text{LSB}]$$

Thus, we can get very good accuracy by repeating the measurement even if the LSB is not small enough [2,7].

5.3 VARIOUS TDC CIRCUITS

There are various TDC circuits that were proposed and developed. The selection of the circuit depends on the required time resolution, hit rate capability, power consumption, and so on.

5.3.1 Counter and Shift Register

Most simple TDC implementations are a counter, as shown in Figure 5.8a, where the counter is started and stopped with an input signal [8]. Normally, this circuit is used to detect only one time interval. To detect multiple transitions of the input signal in a very short time range, a shift register is used (Figure 5.8b) [9]. The drawback of this kind of system is that its time resolution is limited by the clock speed (normally less than 1 GHz, thus the resolution is >1 ns). To have a good time resolution, a very high speed device must be used so that it consumes high power.

5.3.2 Time-to-Amplitude Converter

A precise but simple TDC system, which is used widely in accelerator experiments, is the time-to-amplitude converter (TAC) plus an ADC system (Figure 5.9). The constant current source is connected to a capacitor through a switch that is controlled by an input signal. After switch-off, the charge-up voltage of the capacitor is converted to a digital number by using an ADC. In this system, a time resolution of around 50 ps is achieved. To contain multihit capability, an ASIC chip that contains several TAC channels was developed [10].

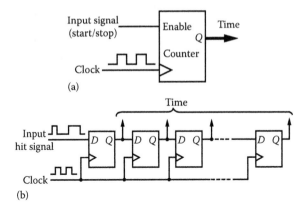

FIGURE 5.8 (a) Time measurement with a counter. (b) TDC by using a shift register to measure the multiple transitions of an input signal.

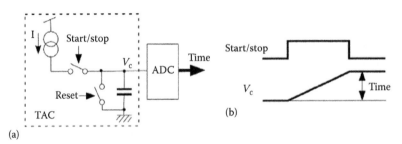

FIGURE 5.9 (a) TAC and ADC system. (b) Timing diagram of the system.

5.3.3 DELAY-LINE TDC

Although the TAC system can achieve fairly good time resolution, it is hard to use in high-rate experiments. Furthermore, it is based on analog circuit, and it needs additional ADC chips. Thus, many kinds of TDC chips that directly output digital value were developed [11].

To have fine time resolution, the signal propagation delay of a buffer is used, as shown in Figure 5.10. A clock or hit signal is passing through a delay line that is composed of buffers, and taps from the buffers are connected to flip-flops. Thus, the transition of the hit signal is recorded in the flip-flops with sequential timing. If the buffer delay is 1 ns and a 32-series buffer is used, the necessary clock period is only 32 ns (~30 MHz), so power consumption is very much reduced compared with the TDC that is implemented with a shift register.

The delay line can be used in a clock signal or hit signal. By delaying the clock signal, a large current flow at a time is avoided, and this scheme will fit to PLL or delay-locked loop (DLL) circuit that will be shown in Figure 5.11. By delaying the hit signal, the output data are available in synchronization to the clock, so it fits to synchronous data acquisition system nicely.

The output of the delay line circuit is a so-called thermometer code like "0001111…" This code is encoded to a binary number by detecting the transition point.

Since the delay time of a buffer will change depending on the process, voltage, and temperature (PVT), the TAC system must be calibrated occasionally to get an accurate value. To stabilize the buffer delay continuously, DLL or PLL circuits are introduced in the digital TDC.

In a DLL system (Figure 5.11a), a voltage-controlled delay element is used instead of a simple buffer. The total delay of the delay line is compared to the clock period and fed back to change the buffer delay. Therefore, the delay is kept constant irrelevant to the PVT change within a controllable range.

In a PLL system, a voltage-controlled ring oscillator (RO) composed of inverters is used. The frequency and phase of the RO are compared with an external clock period and fed back to the RO frequency.

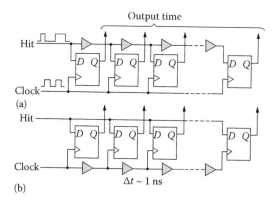

FIGURE 5.10 Delay-line based TDC: (a) hit signal delay and (b) clock delay scheme.

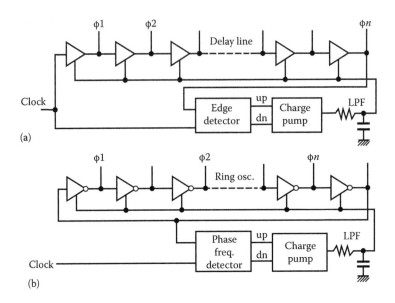

FIGURE 5.11   Basic structure of (a) a DLL circuit and (b) a PLL circuit.

If the spread of the propagation delay is σ and the delay or the period of 2^N elements is calibrated to the clock period, the maximum deviation will be $\sigma\sqrt{2^{N-1}}$ at the center.

There are many discussions on whether DLL or PLL is better for TDC [12–14], but it is difficult to draw a simple conclusion. The DLL circuit is simpler than the PLL, but the layout must be carefully drawn to match the line length of the clock and the delay line signal. In a PLL circuit, the compared value is frequency and not delay, so the line length match is not severe.

In a DLL, if the clock period fluctuates, the measured values are directly affected. On the other hand, in a PLL, the fluctuation of the clock period is filtered in the RO; thus, the effect is mild but continues in a longer time. In general, the DLL scheme is better if the quality of the clock is very good. Otherwise, the PLL scheme may be better.

In a PLL circuit, the RO must have an odd number of inverters to keep the oscillation. This is not preferable, since we would like to have a binary number of taps such as 8, 16, and 32. These values are in even number. Differential buffers connected, as shown in Figure 5.12a, can oscillate in an even number of stages. However, the delay of the differential buffer is normally larger than a single-end buffer.

To get an even number of timing tap from single-end buffers, the asymmetric RO was invented [15]. The rising and falling edges of the oscillating signal pass a different number of inverters. An example of eight timing taps is shown in Figure 5.12b. In this example, the signal transition will pass seven inverters in the first-half period and nine inverters in the second-half period. In total, the signal transitions occur in 16 inverters, and eight equivalent time taps are obtained while keeping the oscillation.

The delay-line TDC circuit is mostly combined with a counter to extend the time range (Figure 5.13). The fine and coarse time values are latched in the next stage

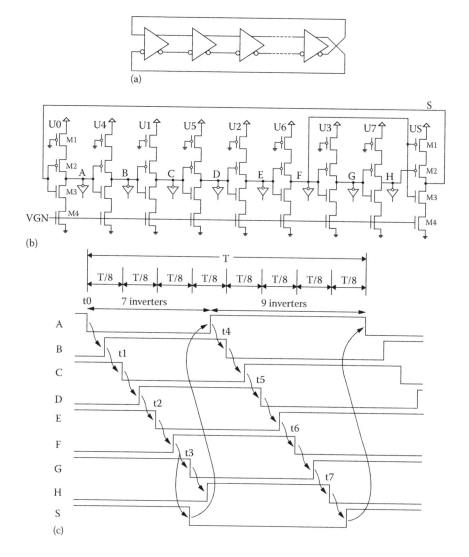

FIGURE 5.12 Even-number stage ROs. (a) RO using differential buffers. (b) Asymmetric RO and (c) its wave form (example of eight stages).

buffer. Since the hit signal is asynchronous to the system clock, the designer must be careful of metastability or glitches when the counter value is latched. One of the solutions to this metastability is to use two latches that are clocked with half-cycle shift. The right counter value is selected by using the fine time value in a later stage.

5.3.4 PULSE-SHRINKING TDC

The pulse-shrinking TDC is an interesting circuit idea that achieves both low-power and gate-delay resolution [16]. Figure 5.14a is a basic delay element that has

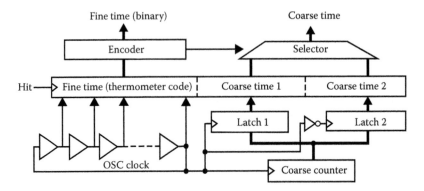

FIGURE 5.13 Delay-line TDC with a coarse counter to get extended time range. To avoid metastability of counter value, two latches clocked with half-cycle shift are used, and the right value is selected by the fine time.

a pulse-width shrink circuit and a pulse-extinction position detection circuit. The delay-line circuit is arranged as an array like the one that is shown in Figure 5.14b. A pulse, of which the width is equal to the time difference between the start and stop, propagates the delay line. In each stage, the pulse width shrinks at a fixed value that is controlled with the "$V_{control}$" signal. Then, the extinction position is detected from the X and Y thermometer codes. This circuit does not need a high-speed clock, and the circuit works only when the input pulse is generated. This is used in TDC circuit in satellites due to its very low power characteristic.

5.3.5 SUBGATE DELAY RESOLUTION

The time resolution of the delay-line TDC is limited by the gate delay, so many methods to get subgate delay resolution are invented.

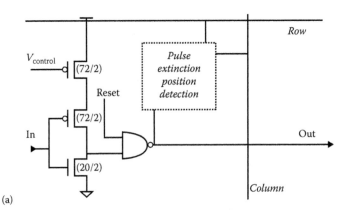

FIGURE 5.14 The pulse-shrinking TDC. (a) A basic element.   (*Continued*)

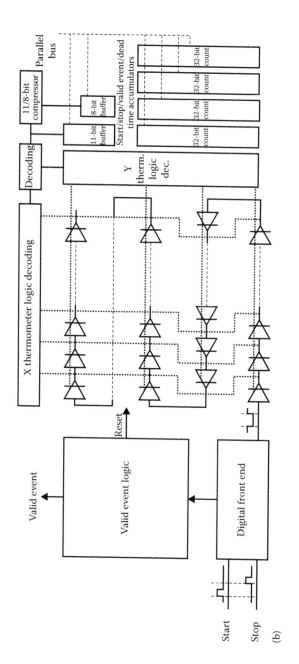

FIGURE 5.14 (CONTINUED) The pulse-shrinking TDC. (b) Configuration of the TDC. (Adapted from K. Karadamoglou et al. 2004. *IEEE J. of Solid-State Cir.*, Vol. 39, No. 1, pp. 214–222.)

5.3.5.1 Phase Interpolator/Blender

To get a shorter delay tap available from the delay buffer, the phase interpolator (or blender) circuit was conceived [17] (Figure 5.15). The signals of ϕ_A and ϕ_B are connected after two inverters and create a ϕ_{AB} signal. By adjusting the PMOS and NMOS drive strength of the inverter, the transition edge of the ϕ_{AB} will be in the middle of ϕ_A and ϕ_B. Repeating the phase blending, a very small delay can be constructed until it is limited by the nonuniformity of the transistor strength.

5.3.5.2 Vernier TDC

By using two different delay lines, we can get finer resolution than the simple delay-line TDC. This method is called a Vernier (or differential-delay) TDC [18] from the resemblance of a Vernier micrometer principle (Figure 5.16). This method is most commonly used to get a subgate delay resolution.

If the delay time of the start delay line is td_1 and that of the stop delay line is td_2, two signals become closer in each stage by Δt ($= td_1 - td_2$). For example, when td_1 is 1 ns and td_2 is 0.9 ns, and the start–stop time difference is 0.6 ns, the stop signal catches up with the start signal at stage 6; thus, we can get a time resolution of $\Delta t = 100$ ps.

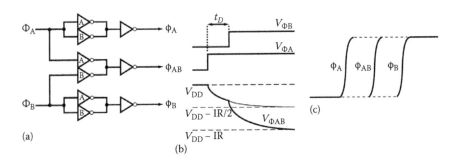

FIGURE 5.15 The phase blender circuit. (a) Basic circuit configuration, (b) signal waveforms, and (c) composed signals. (Adapted from B. W. Garlepp et al. 1999. *IEEE J. of Solid-State Cir.*, Vol. 34, No. 5, pp. 632–644.)

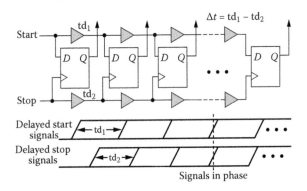

FIGURE 5.16 The Vernier TDC and its timing diagram.

To extend the dynamic range of the vernier TDC, each delay line must have a loop configuration and a loop-counter in each loop. Since the time difference Δt is very short and there must be $N = T_{clock}/\Delta t$ stages, the length of the delay line tends to become very long in Vernier TDC. The long delay line causes a problem of jitter accumulation and mismatches, and the maximum deviation will be at least $\sqrt{2}$ times worse than a single delay-line case.

To avoid the long delay line, two stage sampling schemes using dual PLL [19] or DLL [20] and two-dimension Vernier TDC [21] were developed.

5.3.5.3 Complex RO

By coupling several ROs, an array oscillator is conceived [22] (Figure 5.17a). The output of the dual-input inverter is connected to both horizontal and vertical directions. In steady states, all the output timing will be different. One weak point of this scheme to apply the TDC is the difficulty of extracting timing taps from the center of the array.

To remedy this weak point, the interpolating RO is proposed (Figure 5.17b) [23]. This circuit interpolates between the rise/fall delay times with a single or double pull-down device. In this oscillator, inverters can be arranged as a circular or rectangular shape; thus, it is easy to extract timing taps from the oscillator uniformly.

An example of using the coupled RO is shown in Figure 5.18a [24]. The inner ring and the outer ring are coupled, and the phase of the two rings is synchronized, as shown in Figure 5.18b. The time resolution of 32 ps is achieved for 63 taps. In other developments, even a 1.2-ps resolution is reported [25].

5.3.6 Implementation in an FPGA

Most of the TDCs shown up to here were mostly fabricated in ASICs, but a similar performance was also reported by using an FPGA. Although there are many limitations in FPGA, such as the available number of gates and connections, no analog circuit, and so on, many new ideas are emerging and have achieved a very good time resolution.

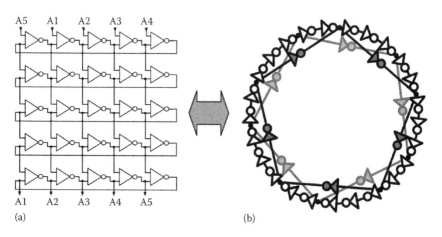

(a)   (b)

FIGURE 5.17   (a) The array oscillator. (b) The coupled RO.

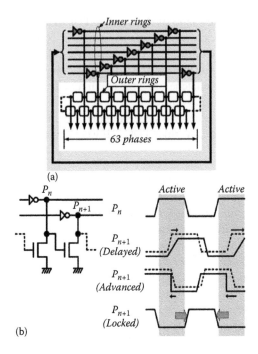

(a)

(b)

FIGURE 5.18 An example of the coupled RO that is used for DVDx16. (a) Seven inner rings and outer rings are coupled to generate 63 phases. (b) Phase-locking scheme. Phases P_n and P_{n+1} are automatically locked.

A naive example of FPGA TDC is shown in Figure 5.19. Many FPGA devices include a DLL circuit to adjust the phase timing of output signals. This can be used to get fine resolution that is smaller than the clock period. In the figure, four-phase clock signals are generated from a 250-MHz base clock with the DLL circuit; then, these signals are used to get a 1-ns resolution.

To get a much better resolution, the delay-line TDC circuit is often implemented in an FPGA. Since it is impossible to implement variable delay elements in an FPGA,

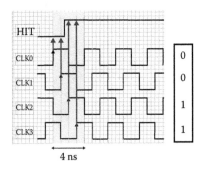

FIGURE 5.19 An example of FPGA TDC, which uses a 250-MHz clock and achieves a 1-ns resolution.

the measured timing data must be calibrated. In addition, available resources are not uniformly located within the FPGA; DNL often becomes worse than ASIC TDC. To remedy the nonuniformity, a multiple time measurement method is invented, and it is called the *wave union* method (Figure 5.20) [26,27]. It generates a pulse train for a single-hit input and measures multiple-edge timing. With this method, the nonuniformity of the delay elements is averaged, and better DNL is achieved.

By measuring temperature and carrying out bin-by-bin calibrations, many FPGA TDCs achieved less than 50-ps resolution in Cyclone II FPGA [28], Virtex-II Pro [29], Virtex-6 FPGA [30,31], etc.

5.3.7 CUSTOM TDC CHIPS

Many TDC ASICs are designed and used in high-energy accelerator experiments. One of the earliest chips was developed in 1988 by using a 0.8-μm CMOS process [32,33] (Figure 5.21). This chip combines a delay element and a memory in a cell and is arranged like a static memory. This chip is named the time memory cell (TMC). All the signal transitions are stored in the memory with 1-ns/bit resolution. The chip includes four blocks of 1-kbit TMC cell and can be used as 1 ch × 4 kbit or 4 ch × 1 kbit. Originally, this chip was developed for the Superconducting Super Collider (SSC) project. Although the SSC was cancelled, TMC chips were used in many experiments such as H1 experiments [34] in DESY laboratory (Germany), D0 in Fermilab (USA), PHENIX [35] experiment in BNL (USA), moon satellite SELENE [36] by JAXA/ISAS (Japan), and Super-Kamiokande [37] by ICRR (Japan).

After the success of the TMC circuit, a much more complex chip was developed for the muon detector of the ATLAS experiment at CERN (Swiss) (called ATLAS muon time-to-digital converter [AMT], Figure 5.22) in collaboration with KEK (Japan) and CERN microelectronics groups [38–40]. To include 24 channels in a

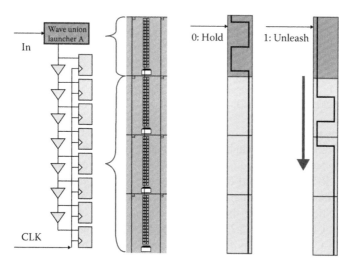

FIGURE 5.20 The wave union method. A pulse train is unleashed to a delay line by receiving an input signal, and multiple-edge timings are measured and averaged.

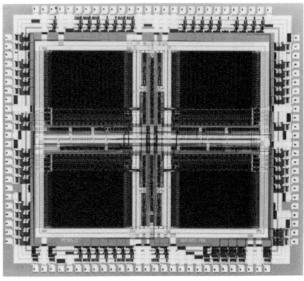

(a)

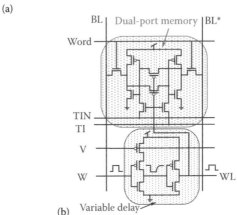

(b)

FIGURE 5.21 (a) Microphotograph of the first TMC chip (TMC1004) fabricated in 0.8-μm CMOS technology in 1990. The size of the chip is 5.0 × 5.6 mm. (b) Basic cell schematic.

chip and long temporary buffers, the event-driven method is taken in the architecture of the AMT chip. It also includes a trigger-matching circuit and serial input/output circuits. The technology used is a 0.35-μm CMOS gate array. Although the transistor size and location are fixed in a gate array, special treatment of manual layout is applied for the RO and PLL parts to get fine resolution and a good DNL/INL.

After the development of the AMT chip, many kinds of TDC chips were also developed in CERN (Figure 5.23) [41,42]. The hit signal is delayed by using resistor-capacitor (R-C) delay chains for four sets of delay-line latches, thus realizing a 25-ps/bit resolution with careful adjustment of the R-C delay. Furthermore, the CERN group recently achieved a 3-ps RMS single-shot precision [43] with 130-nm technology.

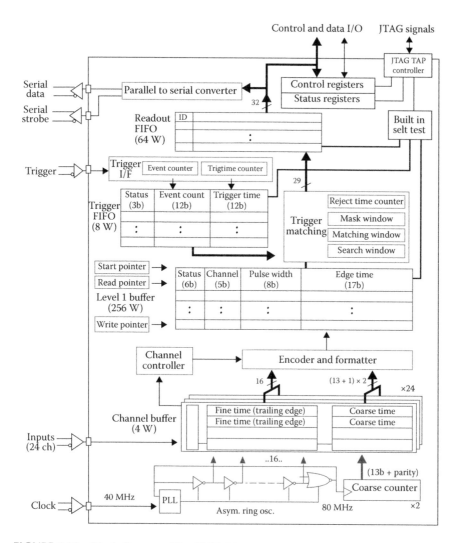

FIGURE 5.22 Block diagram of the AMT chip.

5.4 TDC APPLICATIONS

Recent advances in semiconductor processes make it difficult to design the analog circuit due to the lower operating voltage. Instead of using amplitude domain, many time domain circuits are proposed and used such as all-digital PLLs, sensor interface circuits, TDC-based ADCs, etc. Here, several examples are shown.

5.4.1 COLUMN-PARALLEL ADC IN IMAGE SENSOR

In CMOS image sensors, column-parallel ADCs are often implemented to speed up analog-to-digital (AD) conversion within the chip. In each column, only a comparator

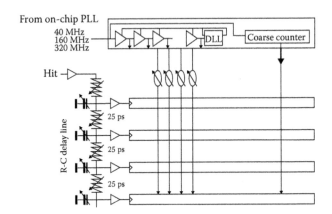

FIGURE 5.23 R-C delay line interpolation in the high-performance TDC. (Adapted from J. Christiansen et al. 2000. A data driven high resolution time-to-digital converter. *Proceedings of the 6th Workshop on Electronics for LHC Experiments*, Krakow, Poland, pp. 169–173, October 11–15, 2000.)

and latch circuits are required. Sensor signals are compared with a common ramp signal, and the time to cross the signals is detected, and a counter value is latched. Conversion time is limited by the counter clock speed. To shorten the conversion time without decreasing the dynamic range of the AD conversion, delay-line TDC is introduced, as shown in Figure 5.24 [44,45].

5.4.2 ADC/TDC FOR AUTOMOBILE

Figure 5.25 shows an all-digital ADC scheme using a delay-line TDC, which is called time analog-to-digital converter (TAD) [46] and developed for automobile application. In addition to its normal operation as a TDC, it can be used as an ADC when an external input voltage is connected to the control voltage of the delay element. The number of counts for a fixed interval correspond to the A/D conversion value. The TAD gives superior environmental resistance and reliability and has a high, inherent adaptability to low-voltage drive. In addition, the A/D conversion resolutions and conversion rates can be controlled by setting the clock period T_s for a variety of applications.

5.4.3 PIXEL TDC

5.4.3.1 Single-Photon Avalanche Diode Detector

A single-photon avalanche diode (SPAD) detector is often required to measure hit timing in pixel array. For example, the time of arrival (ToA) of individual photons with high resolution is needed in several applications, such as fluorescence lifetime imaging microscopy, Förster resonance energy transfer (FRET), optical range finding, and positron emission tomography. In FRET, for example, a typical

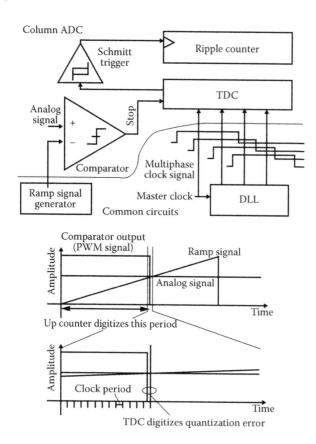

FIGURE 5.24 TDC used in a column-parallel single-slope ADC. Signal crossing time to the master clock edge is measured with a delay-line TDC. (Adapted from M. Shin et al. 2010. Column parallel single-slope ADC with time to digital converter for CMOS imager. *IEEE International Conference on Electronics, Circuits and Systems* [ICECS], Athens, Greece, pp. 863–866, December 12–15, 2010.)

fluorescence lifetime is of the order of 100–300 ps; thus, deep subnanosecond resolutions are needed in the instrument response function. Figure 5.26 shows a pixel schematic for a SPAD image sensor [47]. It contains a 3-bit fine time RO and a 7-bit ripple counter. The resolution of the TDC is 55 ps, and its full scale is 55 ns.

5.4.3.2 Medipix and Timepix Detectors

Hybrid pixel detectors were developed at CERN based on silicon pixel detector technology for high-energy physics experiments [48]. One of these detectors, called Medipix, is a counting-type pixel detector and used in medical and many other applications. Medipix chips measure the pulse height in the time-over-threshold (ToT) method.

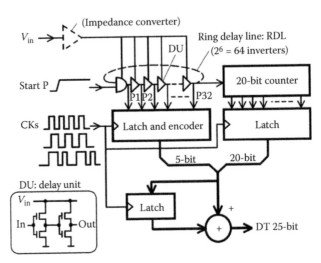

FIGURE 5.25 Block diagram of the TAD. Input signal (V_{in}) is connected to the delay control voltage. (Adapted from T. Watanabe and T. Terasawa. 2009. An all-digital ADC/TDC for sensor interface with TAD architecture in 0.18-μm digital CMOS. *IEEE International Conference on Electronics, Circuits, and Systems, ICECS 2009*, Yasmine Hammamet, Tunisia, pp. 219–222, December 13–16, 2009.)

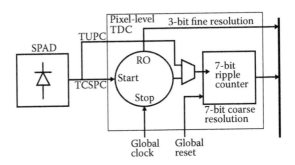

FIGURE 5.26 SPAD pixel schematic: the output of the SPAD acts as a start signal for an RO that is stopped by a global clock. (Adapted from C. Veerappan et al. 2011. A 160 × 128 single-photon image sensor with on-pixel 55 ps 10 b time-to-digital converter. *2011 IEEE Int'l Solid-State Cir. Conf., Digest of Technical Papers*, San Francisco, 17.7, pp. 312–313, February 20–24, 2011.)

As a spin-off of the Medipix chip, Timepix chips were developed, which can also measure time-of-arrival (ToA) in addition to the ToT [49].

5.4.4 All-Digital Phase-Locked Loop

Recently, high-resolution TDC circuits are highlighted in an all-digital phase-locked loop (ADPLL) application [50]. TDC is used to replace the phase/frequency detector in a PLL circuit, as shown in Figure 5.27. This enabled implementing radio frequency (RF) transceivers in an advanced CMOS process and contributes to the progress of mobile devices.

Pioneering work was done by R. B. Staszewski et al. [51]. Normally, in delay-line TDCs, two inverters are used as a delay element to avoid mismatch between rising- and falling-edge transition. In the ADPLL, only one inverter is used as a delay element to get a twice-better resolution (Figure 5.28). It took pseudo-differential topology to avoid the mismatch. In addition, it includes an online calibration circuit to get a stable performance.

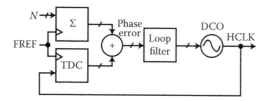

FIGURE 5.27 Block diagram of the ADPLL where TDC is used as a phase/frequency detector. High frequency clock (HCLK) is generated in digitally controlled oscillator (DCO) by referencing the external clock (FREF). (Adapted from R. B. Staszewski et al. 2006. *IEEE Trans. on Cir. and Systems-II: Express Briefs*, Vol. 53, No. 3, pp. 220–224.)

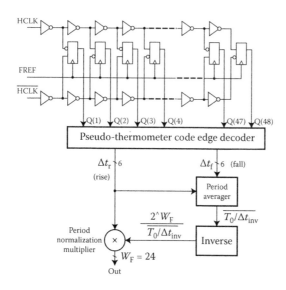

FIGURE 5.28 Pseudo-differential delay-line TDC and online calibration scheme [50] used in the ADPLL. (Adapted from R. B. Staszewski et al. 2006. *IEEE Trans. on Cir. and Systems-II: Express Briefs*, Vol. 53, No. 3, pp. 220–224.)

5.5 ADVANCED TDC CIRCUITS

5.5.1 TIME DIFFERENCE AMPLIFIER

It is impossible to amplify time, but time difference can be amplified. The time difference amplifier (TDA) circuit was proposed in 2002 [52] by utilizing the metastability of mutual exclusion circuit (MUTEX) which output delay is approximately proportional to the input time difference. Then, several different TDA circuits were developed.

Figure 5.29a shows a TDA circuit utilizing two set-reset (SR) flip-flops [53]. If the input time difference ΔT_{SR} between S and R inputs is very small, the metastability of the circuit occurs, and the time to get a stable output level will be delayed. By offsetting the two inputs' timing between A and B by $\pm T_{off}$, the output time difference between A_O and B_O becomes maximum at the offset time. As shown in Figure 5.29c, T_{out} is amplified from T_{in}.

Other smart TDA circuit implementations are shown in Figure 5.30 [54]. In these TDAs, the delay-controlled elements of which delay is changed between a slow and fast state are used. At first, the delay elements are in a fast state. While the two input signals in_1 and in_2 are traveling through a cross-coupled delay chain, they force the delay time of opposite delay element to a slow state. Then, the time difference between outputs out_1 and out_2 will be amplified depending on the input time difference.

The input time range of this TDA can be larger than the previous TDA using metastability, and this can even be cascaded to increase amplifier gain. It is interesting to note that this scheme is very similar to the TDC circuit that was shown in Figure 5.1, which was developed in the 1950s.

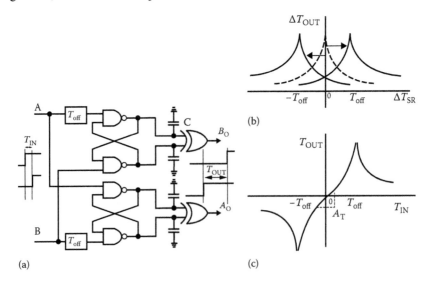

FIGURE 5.29 (a) Concept of a TDA. (b) Shifted SR latch delay characteristics. (c) TDA characteristic. (Adapted from M. Lee and A. A. Abidi. *IEEE J. of Solid-State Cir.*, Vol. 43, No. 4, pp. 769–777.)

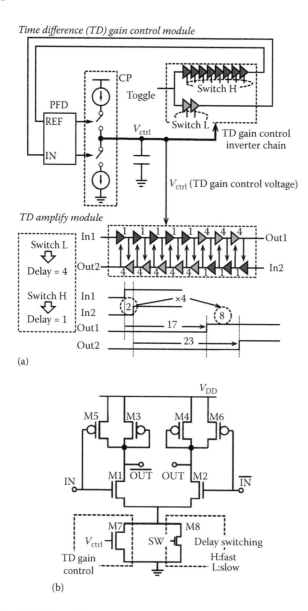

FIGURE 5.30 (a) TDA architecture using closed-loop gain control. (b) Variable delay cell. Delay is controlled both with time difference (TD), gain control voltage (V_{ctrl}), obtained from a charge pump (CP), and delay switching signal (SW). (Adapted from S. Mandai et al. 2010. Cascaded time difference amplifier using differential logic delay cell. *15th Asia and South Pacific Design Automation Conference* [ASP-DAC], Taipei, Taiwan, pp. 355–356, January 18–21, 2010.)

Since time difference can be amplified, it becomes possible to implement a pipelined TDC [55] by using the same kind of architecture that is utilized in pipelined ADC. Figure 5.31 shows a 10-bit pipelined TDC that achieves a 300-MS/s and 1.76-ps resolution.

5.5.2 STOCHASTIC TDC

Time resolution will finally be limited by variation in circuit characteristics in many cases. On the contrary, these variations can be used to get a finer resolution [56]. Instead of connecting a flip-flop to a delay element, several flip-flops are connected to the element (Figure 5.32) [57].

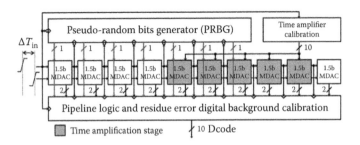

FIGURE 5.31 Architecture of the 10-bit asynchronous pipelined TDC. In each pipeline stage, a multiplying digital-to-analog converter (MDAC) is used. (Adapted from J.-S. Kim et al. 2013. *IEEE J. of Solid-State Cir.*, Vol. 48, No. 2, Feb., pp. 516–526.)

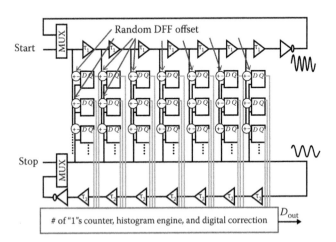

FIGURE 5.32 Stochastic TDC architecture. This figure is showing a self-calibration mode. In a measurement mode, signal is sampled with several D-type flip-flops (DFF) at the same timing, and random offset of the DFF is utilized as time difference. (Adapted from S. Ito et al. 2010. Stochastic TDC architecture with self-calibration. *IEEE Asia Pacific Conference on Circuits and Systems* [APCCAS], Kuala Lumpur, Malaysia, pp. 1027–1030, December 6–9, 2010.)

At first, self-calibration must be done by running two independent ROs. Since the oscillation frequencies of the two oscillators are different and not synchronized, the number of data stored in a histogram would be equal if the TDC is perfect. Here, an encoder circuit, which counts the number of "1" output from the flip-flop, is used instead of a simple thermometer encoder. Digital correction will be done for actual measurement by using these data. With this self-calibration and digital correction scheme, the buffer delay relative mismatches are compensated, and good linearity and resolution are achieved.

5.6   SUMMARY

Over the past several decades, the TDC circuit has progressed very rapidly. Time resolution becomes better and better from the 1-ns era to near-1-ps resolution. The size and power consumption are dramatically reduced, and many TDC channels are included in a chip. Often, the TDC circuits are combined with sensor electronics and hidden in LSI chips.

The applications of TDC become wider and wider. They are used not only in scientific measurements, such as high-energy physics, but also in medical instruments, automobiles, mobile phones, and many other consumer electronics. Furthermore, the TDC circuit is used as a replacement of an analog circuit in advanced LSI processes.

Research on TDC circuit is very active, and many interesting circuit ideas are still emerging, and we cannot see any performance limitations yet.

ACKNOWLEDGMENTS

First, I thank Professor Taka Kondo who encouraged me to start LSI development in KEK. Without him, I would not be able to start the TDC LSI developments. I learned much about LSIs from people in the Nippon Telegraph and Telephone (NTT) Co. LSI laboratory at Atsugi, Japan, with whom I started doing LSI work. I also thank Professor Masayuki Ikebe (Hokkaido University) and Professor Haruo Kobayashi (Gunma University) for their corrections and advice on this article.

REFERENCES

1. J. Christiansen. 2012. Picosecond stopwatches: The evolution of time-to-digital converters. *IEEE Solid-State Circuits Magazine*, pp. 55–59 , Summer 2012, doi: 10.1109/MSSC.2012.2203189.
2. J. Kalisz. 2004. Review of methods for time interval measurements with picosecond resolution. *Institute of Physics Publishing, Metrologia*, Vol. 41, pp. 17–32.
3. S. Henzler. 2010. *Time-to-Digital Converters*. Dordrecht, Netherlands: Springer, ISBN 978-90-481-8627-3.
4. G. Tanahashi. 1958. Chronotron. *BUTSURI*, Vol. 13, No. 2, pp. 72–74. Available at http://dx.doi.org/10.11316/butsuri1946.13.72.
5. Texas Instruments Inc. TDC7200. Available at http://www.ti.com/product/tdc7200.
6. acam messelek gmbh. Available at http://www.acam.de/products/time-to-digital-converters/.

7. Time Interval Averaging, HP application note 162-1. Available at http://www.hpmemory .org/an/pdf/an_162-1.pdf.

8. W. Farr and D. Weiskat. 1981. A low cost multi-hit time to digital converter system for drift chamber. *Nucl. Instr. and Meth.*, Vol. 190, pp. 35–39.

9. O. Sasaki et al. 1989. 1.2 GHz GaAs shift register IC for dead-time-less TDC application. *IEEE Trans. on Nucl. Sci.*, Vol. 36, pp. 512–516.

10. A. E. Stevens et al. 1989. A fast low-power time-to-voltage converter for high luminosity collider detectors. *IEEE Trans. on Nucl. Sci.*, Vol. 36, pp. 517–521.

11. M. S. Gorbics et al. 1997. A high resolution multihit time to digital converter integrated circuit. *IEEE Trans. on Nucl. Sci.*, Vol. 44, No. 3, pp. 379–384.

12. B. Kim et al. 1994. PLL/DLL system noise analysis for low jitter clock synthesizer design. *1994 IEEE International Symposium on Circuits and Systems, ISCAS 94*, Vol. 4, pp. 31–34.

13. L. J. Cheng and Q. Y. Lin. 2001. The performance comparison between DLL and PLL based RF CMOS oscillation. *Proc. 4th International Conference on ASIC*, pp. 827–830.

14. R. C. H. van de Beek et al. 2002. Low-jitter clock multiplication: A comparison between PLLs and DLLs. *IEEE Trans on Cir. and Sys.—II*, Vol. 49, No. 8, pp. 555–566.

15. Y. Arai and M. Ikeno. 1996. A time digitizer CMOS gate-array with a 250 ps time resolution. *IEEE J. of Solid-State Cir.*, Vol. 31, No. 2, pp. 212–220.

16. K. Karadamoglou et al. 2004. An 11-bit high-resolution and adjustable-range CMOS time-to-digital converter for space science instruments. *IEEE J. of Solid-State Cir.*, Vol. 39, No. 1, pp. 214–222.

17. B. W. Garlepp et al. 1999. A portable digital DLL for high-speed CMOS interface circuits. *IEEE J. of Solid-State Cir.*, Vol. 34, No. 5, pp. 632–644.

18. J. F. Genat. 1992. High resolution time-to-digital converters. *Nucl. Instr. and Meth. A* Vol. 315, pp. 411–414.

19. Y. Arai. 2004. A high-resolution time digitizer utilizing dual PLL circuits. *2004 IEEE Nuclear Science Symposium*, Rome, Italy, October 16–22, 2004, conference record N18-3.

20. C.-S. Hwang et al. 2004. A high-precision time-to-digital converter using a two-level conversion scheme. *IEEE Trans. on Nucl. Sci.*, Vol. 51, No. 4, pp. 1349–1352.

21. L. Vercesi et al. 2010. Two-dimensions Vernier time-to-digital converter. *IEEE J. of Solid-State Cir.*, Vol. 45, No. 8, pp. 1504–1512.

22. J. G. Maneatis and M. A. Horowitz. 1993. Precise delay generation using coupled oscillators. *IEEE J. of Solid-State Cir.*, Vol. 29, No. 12, pp. 1273–1282.

23. H. Fadi et al. 2005. A 4.0GHz 0.18μm CMOS PLL based on an interpolative oscillator. *Symp. on VLSI Circuits*, Kyoto, Japan, pp. 100–103, June 14–18, 2005..

24. S. Dosho et al. 2006. A PLL for a DVD-16 write system with 63 output phases and 32 ps resolution. *IEEE International Solid-State Circuits Conference, Digest of Technical Papers*, San Francisco, pp. 2422–2431, February 2006.

25. M. Z. Straayer and M. H. Perrott. 2009. A multi-path gated ring oscillator TDC with first-order noise shaping. *IEEE J. of Solid-State Cir.*, Vol. 44, No. 4, pp. 1089–1098.

26. J. Wu. 2009. An FPGA wave union TDC for time-of-flight applications. *IEEE Nucl. Sci. Symp. Conference Record*, Orlando, pp. 299–304, October 24–November 1, 2009.

27. S.-Y. Wang, J. Wu et al. 2014. A field-programmable gate array (FPGA) TDC for the Fermilab SeaQuest (E906) experiment and its test with a novel external wave union launcher. *IEEE Trans. on Nucl. Sci.*, Vol. 61, No. 6, pp. 3592–3598.

28. W. Pan, G. Gong and J. Li. 2014. A 20-ps time-to-digital converter (TDC) implemented in field-programmable gate array (FPGA) with automatic temperature correction. *IEEE Trans. on Nucl. Sci.*, Vol. 61, No. 3, pp. 1468–1473.

29. M. A. Daigneault and J. P. David. 2011. A high-resolution time-to-digital converter on FPGA using dynamic reconfiguration. *IEEE Trans. on Instr. and Meas.*, Vol. 60, No. 6, pp. 2070–2079.

30. H. Menninga et al. 2011. A multi-channel, 10 ps resolution, FPGA-based TDC with 300 MS/s throughput for open-source PET applications. *IEEE NSS/MIC*, Valencia, Spain, pp. 1515–1522, October 23–29, 2011..

31. M. W. Fishburn et al. 2013. 19.6 ps, FPGA-based TDC with multiple channels for open source applications. *IEEE Trans. on Nucl. Sci.*, Vol. 60, pp. 2203–2208.

32. Y. Arai and T. Baba. 1988. A CMOS time to digital converter VLSI for high-energy physics. *Symposium on VLSI Circuits*, Tokyo, IEEE CAT. No. 88, TH 0227-9, pp. 121–122, August 22–24, 1988.

33. Y. Arai, T. Matsumura and K. Endo. 1992. A CMOS 4 ch × 1 k Time Memory LSI with 1 ns/bit Resolution. *IEEE J. of Solid-State Cir.*, Vol. 27, No. 3, pp. 359–364.

34. E. Eisenhandler et al. 1995. The H1 SPACAL Time-to-Digital Converter System. *IEEE Trans. on Nucl. Sci.*, Vol. 42, No. 4, pp. 688–692.

35. Y. Arai et al. 1998. Time memory cell VLSI for the PHENIX drift chamber. *IEEE Trans. on Nucl. Sci.*, Vol. 45, No. 3, pp. 735–739.

36. Y. Saito et al. 2010. In-flight performance and initial results of Plasma Energy Angle and Composition Experiment (PACE) on SELENE (Kaguya). *Space Sci. Rev.*, Vol. 154, pp. 265–303.

37. S. Yamada et al. 2010. Commissioning of the new electronics and online system for the Super-Kamiokande Experiment. *IEEE Trans. on Nucl. Sci.*, Vol. 57, pp. 428–432.

38. Y. Arai. 2000. Development of front end electronics and TDC LSI for the ATLAS MDT. *Nucl. Instr. Meth. A* Vol. 453, pp. 365–371.

39. Y. Arai et al. 2004. On-chamber readout system for the ATLAS MDT Muon Spectrometer. *IEEE Trans. on Nucl. Sci.*, Vol. 51, pp. 2196–2200.

40. ATLAS TDC home page. Available at http://research.kek.jp/group/atlas/tdc/.

41. J. Christiansen et al. 2000. A data driven high resolution time-to-digital converter. *Proceedings of the 6th Workshop on Electronics for LHC Experiments*, Krakow, Poland, pp. 169–173, October 11–15, 2000.

42. M. Mota et al. 2000. A flexible multi-channel high-resolution time-to-digital converter ASIC. *IEEE Nucl. Sci. Sympo. Conference Record*, Vol. 2, pp. 9/155–9/159.

43. L. Perktold and J. Christiansen. 2013. A fine time-resolution (<3 ps-rms) time-to-digital converter for highly integrated designs. *2013 IEEE International Instrumentation and Measurement Technology Conference* (I2MTC), Minneapolis, pp. 1092–1097, May 6–9, 2013.

44. M. Shin, M. Ikebe, J. Motohashi and E. Sano. 2010. Column parallel single-slope ADC with time to digital converter for CMOS imager. *IEEE International Conference on Electronics, Circuits and Systems* (ICECS), Athens, Greece, pp. 863–866, December 12–15, 2010.

45. K. Kim, M. Ikebe and J. Motohashi. 2012. A 11b 5.1 μW multi-slope ADC with a TDC using multi-phase clock signals. *IEEE International Conference on Electronics, Circuits and Systems* (ICECS), Seville, Spain, pp. 512–515, December 9–12, 2012.

46. T. Watanabe and T. Terasawa. 2009. An all-digital ADC/TDC for sensor interface with TAD architecture in 0.18-μm digital CMOS. *IEEE International Conference on Electronics, Circuits, and Systems, ICECS 2009*, Yasmine Hammamet, Tunisia, pp. 219–222, December 13–16, 2009.

47. C. Veerappan et al. 2011. A 160×128 single-photon image sensor with on-pixel 55 ps 10 b time-to-digital converter. *2011 IEEE Int'l Solid-State Cir. Conf., Digest of Technical Papers*, San Francisco, 17.7, pp. 312–313, February 20–24, 2011.

48. M. Campbell et al. 1998. Readout for a 64 × 64 pixel matrix with 15-bit single photon counting. *IEEE Trans. on Nucl. Sci.*, Vol. 45, No. 3, pp. 751–753.

49. T. Poikela et al. 2014. Timepix3: A 65K channel hybrid pixel readout chip with simultaneous ToA/ToT and sparse readout. *JINST* Vol. 9, pp. C05013. doi:10.1088/1748-0221/9/05/C05013.

50. R. B. Staszewski et al. 2006. 1.3 V 20 ps time-to-digital converter for frequency synthesis in 90-nm CMOS. *IEEE Trans. on Cir. and Systems-II: Express Briefs*, Vol. 53, No. 3, pp. 220–224.

51. R. B. Staszewski et al. 2005. All-digital PLL and transmitter for mobile phones. *IEEE J. of Solid-State Cir.*, Vol. 40, pp. 2469–2482.

52. A. M. Abas et al. 2002. Time difference amplifier. *Electronics Letters*, Vol. 38, No. 23, pp. 1437–1438.

53. M. Lee and A. A. Abidi. A9 b, 1.25 ps resolution coarse–fine time-to-digital converter in 90 nm CMOS that amplifies a time residue. *IEEE J. of Solid-State Cir.*, Vol. 43, No. 4, pp. 769–777.

54. S. Mandai et al. 2010. Cascaded time difference amplifier using differential logic delay cell. *15th Asia and South Pacific Design Automation Conference* (ASP-DAC), Taipei, Taiwan, pp. 355–356, January 18–21, 2010.

55. J.-S. Kim et al. 2013. A 300-MS/s, 1.76-ps-resolution, 10-b asynchronous pipelined time-to-digital converter with on-chip digital background calibration in 0.13-μm CMOS. *IEEE J. of Solid-State Cir.*, Vol. 48, No. 2, Feb., pp. 516–526.

56. P. M. Levine and G. W. Roberts. 2004. A high-resolution flash time-to-digital converter and calibration scheme. *Proceedings of International Test Conference*, Charlotte, pp. 1148–1157, October 26–28, 2004.

57. S. Ito et al. 2010. Stochastic TDC architecture with self-calibration. *IEEE Asia Pacific Conference on Circuits and Systems* (APCCAS), Kuala Lumpur, Malaysia, pp. 1027–1030, December 6–9, 2010.

6 Digital Pulse-Processing Techniques for X-Ray and Gamma-Ray Semiconductor Detectors

Leonardo Abbene, Gaetano Gerardi, and Fabio Principato

CONTENTS

6.1   Introduction 121
6.2   Pulse-Processing Electronics: Analog and Digital Approaches................. 122
6.3   A New Real-Time DPP System for X- and Gamma-Ray Semiconductor
      Detectors.. 126
      6.3.1   DPP Firmware 127
      6.3.2   Pulse Detection.. 127
      6.3.3   PSHA 127
      6.3.4   ICR, OCR, and Dead Time.. 129
      6.3.5   Working Modes 130
6.4   Performance of the DPP System with Semiconductor Detectors............... 131
      6.4.1   Measurements with a CdTe Detector.. 132
      6.4.2   Measurements with a Ge Detector 135
References.. 138

6.1   INTRODUCTION

Over the last decade, digital pulse-processing (DPP) electronics have been widely proposed and used for new generation x- and gamma-ray spectrometers. DPP systems, based on direct digitizing and processing of detector signals, lead to better results than the traditional analog pulse-processing electronics in terms of stability, flexibility, reproducibility, energy resolution, throughput, and dead time. In this chapter, we will review the principles of operation of conventional analog electronic chains for x- and gamma-ray semiconductor detectors, with special emphasis on the benefits of the digital approach. The characteristics of a new real-time DPP system, developed by our group, are discussed in depth. Finally, we present some original results on cadmium telluride (CdTe) and germanium (Ge) detectors, highlighting the excellent performance of the DPP system both at low and high counting rate environments (up to 1.1 Mcps).

6.2 PULSE-PROCESSING ELECTRONICS: ANALOG AND DIGITAL APPROACHES

The typical electronic chain for x- and gamma-ray semiconductor detectors is shown in Figure 6.1. The first element is the charge-sensitive preamplifier (CSP), which provides the interface between the detector and the following pulse-processing electronics (shaping amplifier and multichannel analyzer [MCA]). The preamplifier is designed to integrate the induced current arising from the movement of electrons and holes in the detector under the influence of an applied electric field. It converts the current pulses into voltage pulses, generally characterized by a fast exponential leading edge followed by a slow exponential decay (resistive-feedback preamplifiers). The leading edge, generally described by the peaking time (i.e., the time at which the pulse reaches its maximum), depends on the time width of the current pulse, i.e., on the collecting time of the charge that is created in the detector (typical values of a few hundreds of nanoseconds). The pulse height is proportional to the generated charge and then to the energy of the detected events. The exponential falling edge decreases with a time constant (typically a few hundreds of microseconds) that is equal to the product of the capacitance and resistance of the feedback loop (resistive-feedback preamplifiers). As is well known, there are events releasing the same energy but with different charge collection times (typical of semiconductor detectors). To avoid these events that can produce pulses with different heights, the time constant is typically chosen several orders of magnitude greater than the peaking time. On the other hand, a time constant too large can produce overlapped preamplified pulses. This effect can be partially reduced by shortening the width of each preamplified pulse without altering its height (pulse-shaping techniques). The primary functions of the shaping amplifier (linear amplifier) are to shorten and to amplify the CSP output pulses. The simplest concept for a pulse-shaping amplifier is the use of a capacitor-resistor (CR) filter (high-pass filter in the frequency domain or differentiator in the time domain) to shorten the pulses, followed by a resistor-capacitor (RC) filter (low-pass filter in the domain of frequencies or integrator in the time domain) to improve the signal/noise ratio. The differentiation of the CSP output pulses produces a pulse undershoot due to the finite decay time of the CSP output pulses. Pole-zero cancellation techniques are usually applied to eliminate this undershoot [1]. At high photon-counting rates, poor compensation of undershoot creates a random baseline shift that reduces the precision of the pulse height measurements. This effect can be partially reduced

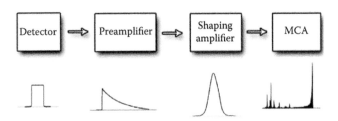

FIGURE 6.1 Schematic block diagram of a typical detection system for x- and gamma-ray spectroscopy. Typical pulse shapes produced by each element and the final result.

by using bipolar pulses or sophisticated techniques that are able to compensate for the random baseline shift (baseline restorer techniques) [1]. Therefore, good shaping amplifiers must be equipped with pole-zero cancellation and baseline restorer. The shaping amplifiers are generally analog devices that are characterized by different pulse shapes (CR-RC, Gaussian, triangular, trapezoidal) and a few different shaping time constant values (generally approximately six values). Finally, the last stage of the processing chain concerns the measurement of the shaped pulse heights (i.e., the energy of the events) and the creation of the pulse height spectra (energy spectra). This process is known as an MCA mode. An MCA is used to sample and to record the shaped pulse heights and to generate the energy spectrum. Generally, the main task of a pulse-processing chain is to give the best energy resolution possible and the input counting rate (ICR). Moreover, it is also very important to obtain energy spectra with a good counting statistics, i.e., characterized by a sufficient number of counts in the measured spectra. Under specific experimental conditions, in which the acquisition time is limited (typical in medical applications), the output counting rate (OCR), i.e., the measured counting rate in the spectrum, has a key role. This characteristic is generally given by the throughput of the processing chain, typically referred to as the OCR that is related to the ICR. Generally, the shaping time constant of a shaping amplifier is an important parameter for both optimum energy resolution and throughput. The choice of the proper value of the shaping time constant (e.g., the time constant of the CR-RC filters) should take into account several factors: (a) ballistic deficit, (b) noise, (c) pileup, and (d) dead time [1]. To preserve the height of the CSP pulses, the shaping time constant must be large compared to the peaking time. The loss of height, due to a finite shaping time constant, is referred to as ballistic deficit. As well outlined in the literature [1,2], noise can be categorized into series and parallel sources. Series noise sources are mainly due to the thermal noise of the input field-effect transistor (FET) of the preamplifier, whereas parallel noise includes the fluctuations of detector and FET leakage currents and the thermal noise of the feedback resistance. Pileup phenomena occur when the shaped pulses overlap on the tail (undershoot or overshoot) from a preceding pulse (tail pileup) or when two pulses are sufficiently close together so that they are treated as a single pulse (peak pileup), producing severe degradations in the pulse height spectra (i.e., the energy spectra). Tail pileup generally produces worsening in the energy resolution, distorting the peak shapes of the pulse height spectra, whereas peak pileup adds new peaks that are not related to true events. The dead time, i.e., the time needed to process one pulse, is mainly due to the detector collection time, the width of the shaped pulses, the MCA conversion time, and the MCA data storage time. The major contributions to the dead time of a detection system are due to the shaping amplifier and the MCA (typically values between a few microseconds and a few tens of microseconds). Knowledge of the dead time of a system is essential for accurate estimations of the true ICR. If the dead time is well known, the true ICR can be obtained through the measured counting rate and by using a proper dead time model. Typically, two kinds of dead time models can be distinguished: (1) paralyzable and (2) nonparalyzable dead time. For a paralyzable model, an event occurring during the dead time belonging to a previous pulse, although it will be lost, still starts a new dead time period extending the overall dead time. For a nonparalyzable model, an event occurring during the dead time is

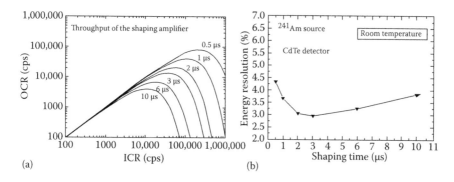

(a)

(b)

FIGURE 6.3 (a) Calculated OCR of a shaping amplifier as a function of the ICR, at different shaping time constant values. (For this amplifier, the dead time is nine times the shaping time constant.) (b) Energy resolution, measured with a CdTe detector, at different shaping time constant values.

have proposed hybrid (i.e., both analog and digital) pulse-processing chains, in which the shaped pulses from an analog amplifier are sampled by a digitizer (with sampling frequencies > 10 MHz), thus eliminating the dead time of MCAs (Figure 6.4a). The digitized shaped pulses are processed offline for pulse height analysis and pileup inspections. These systems showed good spectroscopic performance up to photon-counting rates of approximately 100 kcps, a limit due to the finite width of the shaped pulses and the difficulties on baseline restoration. At higher counting rates, the direct digitization of the preamplifier output pulses (DPP approach) is a very appealing solution, as reported in several works [7–17]. In a DPP system, the preamplifier output (CSP) signals are directly digitized by ADCs (with sampling frequencies ≥ 100 MHz) and then processed by using digital algorithms (Figure 6.4b). A DPP system leads to better results than the analog one in terms of parameters such as stability, flexibility,

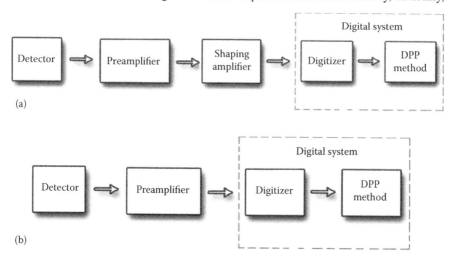

(a)

(b)

FIGURE 6.4 Simplified block diagrams of (a) a hybrid pulse processing chain and (b) a DPP chain.

reproducibility, energy resolution, throughput, dead time modeling, and the possibility of shape preservation of the pulses for further analysis. In a DPP system, the direct digitizing of the detector signals minimizes the drift and instability that are normally associated with analog pulse processing. Once digitized, the pulses are immune to distortions that are caused by electronic noise and temperature instabilities. Moreover, it is possible to use complex algorithms for adaptive processing and optimum filtering, which are not easily implementable in a traditional analog approach. A DPP analysis also requires considerably less overall processing time than the analog ones, ensuring lower dead time and higher throughput. In a DPP system, there is no additional dead time that is associated with digitizing the pulses, and so there is no MCA dead time (conversion time and data storage time). Preservation of the detector pulse shape for pulse shape analysis is also very important for performance enhancements, photon tracking, or particle identification. Some DPP systems are composed of a digitizer and a personal computer for data recording and analysis (*offline analysis*) [3–6,10,12–16]. *Real-time data processing* [7–9,17,18], in which the signals are acquired, recorded, and processed online, is obtained by using digitizers with local memory and field-programmable gate arrays (FPGAs) wherein pulse-processing algorithms can be implemented (DPP firmware).

In Section 6.3, we will report on the characteristics and performance of a new real-time DPP system for x- and gamma-ray semiconductor detectors. The system is based on a modified version of a commercial digitizer that is equipped with a custom DPP firmware, developed by our group, for online pulse shape and height analysis (PSHA). Experimental results on CdTe and Ge detectors, at both low and high photon-counting rates, are reported. Moreover, a parallel comparison with a standard analog pulse-processing electronics, in terms of energy resolution and throughput, is also performed.

6.3   A NEW REAL-TIME DPP SYSTEM FOR X- AND GAMMA-RAY SEMICONDUCTOR DETECTORS

The DPP system is composed of a digitizer and a PC, where the user can set the DPP parameters, choose the working mode, and display the results by using a Labview program, developed by our group (control and display software). Figure 6.5 shows a

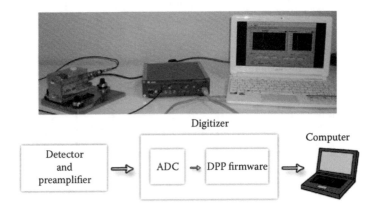

FIGURE 6.5   The DPP system coupled to a CdTe detector.

picture and the block diagram of the digital system that is coupled to a CdTe detector. The pulse-processing analysis is performed by using a custom DPP firmware, developed by our group and uploaded to the digitizer. To digitize the preamplifier output waveform, we used a commercial digitizer (DT5724, CAEN S.p.A., Italy), housing high-speed ADCs (14 bits and 100 MS/s), a buffer memory, and Cyclone EP1C20 FPGAs (ALTERA, USA). The digitizer is equipped with four channels (*AC* coupled). Each channel is characterized by one ADC with three full-scale ranges (±1.125, ±0.5625, and ±0.2813 V). The data stream from each ADC is written in four circular memories without dead time. The DPP is carried out by the dedicated FPGA in which the DPP method is implemented (DPP firmware). The produced data are packed, and another FPGA collects and sends them, via a Universal Serial Bus channel, to the PC.

6.3.1 DPP FIRMWARE

The DPP firmware is based on a revised version of a DPP method, developed by our group and successfully used for both offline [10,12–15] and online analysis [17,18]. The pulse detection is performed by using a *fast*-shaping mode, whereas the PSHA starts with a *slow*-shaping mode.

The DPP firmware is characterized by two main features: (1) it performs the PSHA of each event on single isolated input waveform windows to avoid the corruptions that the analysis can produce to adjacent pulses, and (2) due to an automatic baseline restoration (based on the analysis on single pulses), it also allows AC coupling. This approach allows the elimination of the average value of the preamplifier output waveform and thus the maximum exploitation of the ADC input ranges.

In Sections 6.3.2 through 6.3.5, we will summarize the main operations that are performed by the DPP firmware.

6.3.2 PULSE DETECTION

Once digitized, the preamplifier output waveform is shaped by using the classical single-delay-line (SDL) shaping technique [1]. Shaped pulses are obtained by subtracting from the original waveform its short-delayed and attenuated fraction. The attenuation is able to avoid undesirable undershoots in the shaped pulses, therefore working as the classical pole-zero cancellation technique [1]. The width of each shaped pulse (fast SDL pulses), without noise, is equal to $T_d + T_p$, wherein T_d is the delay time, and T_p is the peaking time of the original pulse. This shaping operation (fast-shaping mode), characterized by a short delay time, is able to detect the pulses (i.e., the events) from the preamplifier output waveform with high time resolution, as shown in Figure 6.6. A trigger signal is generated and time-tagged when the falling edge of the shaped pulses crosses a height threshold, obtained by using the amplitude and rise-time compensating technique [1].

6.3.3 PSHA

A PUR is first performed by the DPP method. The trigger signal will detect a non-piled-up event if there are no other triggers within a selected time window (centered on the trigger signal) of the preamplifier output waveform. The width of the window

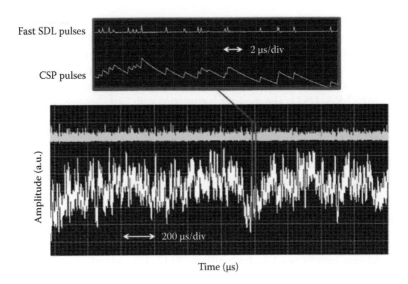

FIGURE 6.6 The digitized preamplifier output (CSP) pulses (white line) and the shaped pulses (cyan line) through the fast-shaping operation. A zoom of the picture clearly shows the good detection of the pulses from the waveform. The pulses represent [241]Am photons impinging on a thin CdTe detector (1-mm thick) with an ICR of 825 kcps.

is termed snapshot time (ST), and all the non-piled-up events are analyzed. In order to avoid the corruptions that the analysis can produce to adjacent pulses, the DPP method performs the PSHA of the selected events on the extracted snapshot waveform, ST wide, centered on the trigger. Figure 6.7a shows an example of a short snapshot waveform. Moreover, three snapshot waveforms, with each maximum amplitude at the center position of the window (ST = 3 μs), at different ICRs, are shown in Figure 6.7b.

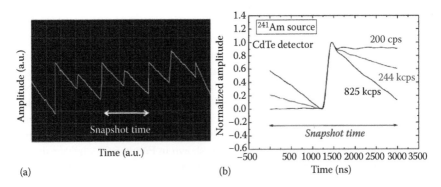

FIGURE 6.7 (a) The snapshot waveform. Each preamplified pulse is presented within a time window that is equal to the selected ST. (b) Single isolated preamplifier output pulses, within an ST = 3 μs, at different ICRs. The waveform decay changes, before and after the leading edge of each pulse, are due to the different number of piled-up pulses. The pulses represent [241]Am photons impinging on a CdTe detector. Each pulse is arbitrary vertically shifted and normalized to its maximum amplitude.

It is clearly visible that the waveform decay, before and after the leading edge of each pulse, depends on the ICR. This effect is due to the piling up of the pulses on the exponential tail of previous pulses, which rises up as the ICR increases. Once extracted, the snapshot waveform is shaped by using a slow SDL operation, characterized by a longer delay time than the fast one. (We used a delay time = ST/4.) The slow SDL shaping is used to perform a baseline restoration, by evaluating the mean value of the samples preceding the leading edge of the shaped pulses, a pulse shape analysis, by evaluating the maximum amplitude and the rise time of the shaped pulses, and the height analysis of the shaped pulses. By using the proper decay constant value, the slow SDL shaping allows the elimination of the influence of the previous events on each selected pulse and, through the baseline recovery, the minimization of baseline shifts (due to AC coupling and DC offsets) that are more severe at high ICRs. The maximum amplitude of the shaped pulses is obtained after the compensation of the exponential decay by using a digital deconvolver [19]. The shape analysis consists of the measurement of the peaking time of the shaped pulses. The rise time of the pulses, i.e., the interval between the times at which the shaped pulse reaches 10% and 90% of its maximum amplitude, is first evaluated. The times, corresponding to the exact fractions (10% and 90%) of the pulse height, are obtained through a linear interpolation. The method estimates the peaking time that is equal to 2.27 times the rise time (i.e., approximately five times the time constant) with a precision of 2 ns.

Finally, the pulse height (event energy) estimation is performed by applying an optimized low-pass filter to all the samples of each shaped and deconvolved pulse.

As is well known [1,12–15], the correlation between the height and the peaking time of the pulses can be generally used to minimize both incomplete charge collection (Figure 6.8a) and peak pileup effects (Figure 6.8b). To compensate for incomplete charge collection, pulse shape discrimination (PSD) and pulse shape correction (PSC) techniques [14,17] were implemented. Moreover, the height–peaking time correlation is also helpful to minimize the peak pileup effects (i.e., overlapped preamplifier output pulses within the peaking time) in the measured spectra. Because the peak pileup pulses are characterized by a longer peaking time and a higher height than the correct pulses, it is possible to reduce their effects in the measured spectra by analyzing both the peaking time and the height (energy) distribution of the pulses [14–17] (Figure 6.8b). We implemented a PSD technique that is based on the selection of the proper peaking time region (PTR).

6.3.4   ICR, OCR, AND DEAD TIME

The well-defined dead time modeling is one of the most interesting characteristics of our DPP system. Knowledge of the dead time and its analytic model is very important in estimating the true ICR. Both the width of each fast-shaped pulse (with no pileup) and the ST are dead times for the DPP system, and they can be used in a paralyzable model (Equation 6.1). Therefore, the DPP system is then always able to estimate the ICR through Equation 6.1, the knowledge of the dead time, and by using the measured OCR_{FAST} from fast shaping. Figure 6.9a shows both the calculated ICR and the measured OCR_{FAST} from fast shaping (dead time = 170 ns) versus the tube current (W-target x-ray source, CdTe detector). OCR_{SLOW} (ST = 5 μs) versus OCR_{FAST} values are shown in Figure 6.9b. The calculated ICR versus the tube current shows nonlinearity < 0.09%

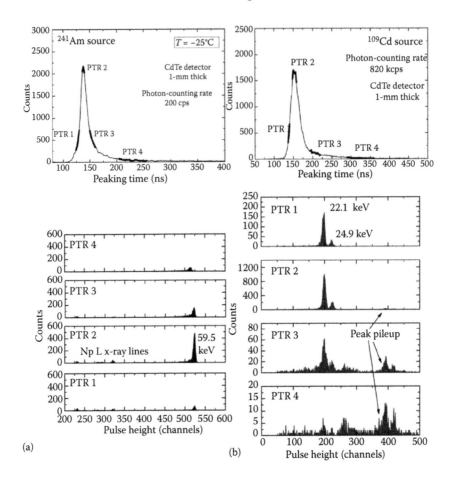

FIGURE 6.8   (a) Peaking time distribution of the pulses (^{241}Am photons at low ICR = 200 cps) on a planar CdTe detector (1-mm thick) and the measured spectra of the pulses that were selected at four peaking time regions (PTRs). (b) Peaking time distribution of ^{109}Cd pulses at high ICR of 820 kcps and the spectra for the selected PTRs. It is well evident that both incomplete charge collection pulses and the peak pileup pulses are characterized by longer peaking times than the correct pulses.

up to 1.1 Mcps. This result is highlighted as our DPP system is always able to estimate the true rate of the impinging photons through fast shaping, even for low rates in the spectrum (slow shaping). Therefore, it is possible to use long ST values (low through-put) for optimum pulse height analysis (i.e., for optimum energy resolution) without perturbing the correct estimation of the input photon-counting rate.

6.3.5   WORKING MODES

To obtain the overall features of the impinging photons, the DPP can transmit one of the following several results to the PC: (a) input waveform; (b) fast-shaping waveform; (c) energy and peaking time list; (d) energy, peaking time, and time of occurrence

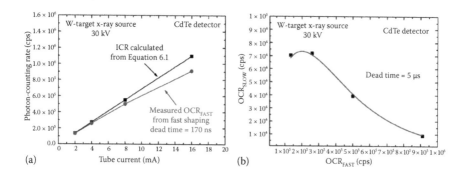

FIGURE 6.9 (a) Measured OCR$_{FAST}$ from fast shaping versus tube current (W target) up to 1.1 Mcps (red points). The calculated ICR from OCR$_{FAST}$ and by using Equation 6.1 shows excellent linearity versus the tube current up to 1.1 Mcps. (Nonlinearity is less than 0.09%.) (b) OCR$_{SLOW}$ versus OCR$_{FAST}$ (ST = 5 μs).

list; (e) time of occurrence and reshaped (after baseline restorer and exponential time decay correction) pulse leading edge list; and (f) the snapshot waveform (a sequence of preamplified pulses with their time of occurrence). In the last working mode, the system only transmits the preamplified pulses that are selected for the pulse height analysis, i.e., not piled up with the preceding and following pulses in the input waveform. Each pulse is presented within a time window that is equal to the selected ST. Contrary to what happens in analog pulse-processing systems or in other DPPs using similar processing operations, the preamplifier pulse shape is preserved. We stress that each data, packed as required from the selected working mode, are coupled to the following housekeeping data: (a) time stamp of the packed data, (b) total accumulated dead time, (c) number of fast-detected pulses, (d) number of analyzed events, and (e) number of pileup events (slow and fast PUR). From these data, the user can derive, with high time resolution, the related count rates. Each working mode is set through the control and display software.

6.4 PERFORMANCE OF THE DPP SYSTEM WITH SEMICONDUCTOR DETECTORS

In Sections 6.4.1 and 6.4.2, we will highlight the performance of the DPP system, in terms of energy resolution and throughput, coupled to a thin, planar CdTe detector and a coaxial Ge detector (p-type). As is well known, thin CdTe/CdZnTe detectors (1–2-mm thick) are very appealing for x-ray spectroscopy in the 1–100-keV energy range [20–30], whereas coaxial p-type Ge detectors are excellent spectrometers in the gamma energy range (>100 keV) [1,31]. We used a planar CdTe detector (XR100T-CdTe, S/N 6012, Amptek, United States) with a thickness of 1 mm (absolute efficiency of 64% at 100 keV) [12] and a coaxial Ge detector (GEM40P4-76, Ortec, United States) with a crystal diameter of 64.1 mm and a crystal length of 64.4 mm (relative efficiency of 40% at 1.33 MeV) [32]. Both detectors are equipped with resistive-feedback CSPs.

For comparison, a standard analog pulse-processing chain was also used. The preamplifier output pulses were shaped by an analog shaping amplifier (672, Ortec, United States), equipped with different shaping time constant values of 0.5, 1, 2, 3, 6, and 10 μs. The semi-Gaussian output pulses were acquired by a standard MCA (MCA-8000A, Amptek, United States).

We measured the system response to four x- and gamma-ray calibration sources (^{109}Cd: 22.1, 24.9, and 88.1 keV; ^{241}Am: 59.5 keV; ^{137}Cs: 661.7 keV; ^{60}Co: 1173.2 and 1332.5 keV). To obtain different impinging photon-counting rates, we changed the irradiated area of the detectors using collimators (Pb and W) with different geometries. A W-target x-ray tube was also used.

6.4.1 MEASUREMENTS WITH A CdTe DETECTOR

We measured x-ray spectra at low and high ICRs by using both the analog and the DPP systems. At low ICR (200 cps), the DPP system shows a similar performance to the analog one. However, the DPP system allows PSD and PSC to compensate for incomplete charge-collection effects in the measured spectra (energy resolution degradation, peak asymmetry, tailing). Figure 6.10 shows ^{57}Co spectra, measured with the CdTe detector, after digital PSD and nonlinear PSC. We used an ST = 40 μs. Despite the strong reduction of the photon counts (93%), PSD produced good improvements in the spectrum: energy resolution of 1.6% *full width half maximum* (*FWHM*) at 122 keV with no correction and of 0.6% FWHM after PSD. Good results were also obtained (0.8% FWHM at 122 keV) by using the nonlinear PSC, with no reduction of photon counts.

Important differences between the analog and digital approaches were well highlighted at high ICRs. We investigated the high-rate performance of the DPP system by analyzing the stability of both peak position (centroid) and energy resolution at various ICRs. We measured ^{109}Cd and ^{241}Am spectra at various ICRs with both systems working at similar throughput. Concerning analog electronics, we used a shaping time constant of 0.5 μs (the smallest selectable value from the analog

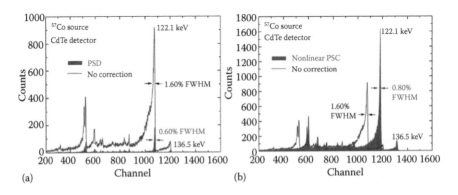

(a)   (b)

FIGURE 6.10 (a) Measured ^{57}Co spectra using PSD and no PSD techniques. After PSD, we obtained an energy resolution of 0.60% FWHM at 122.1 keV. (b) An energy resolution of 0.80% FWHM at 122.1 keV was obtained after nonlinear PSC, with no reduction of photon counts.

amplifier), which is the best value for both maximum throughput and optimum energy resolution at ICRs > 30 kcps. (This result was obtained experimentally by analyzing ^{109}Cd and ^{241}Am spectra at different shaping time constant values and at different ICRs.) To obtain a similar analog throughput, we used an ST of 3.4 and 4.2 μs for ^{109}Cd and ^{241}Am sources, respectively. The measurements are shown in Figure 6.11. The maximum rate range is fixed by the limits of the analog pulse-processing electronics (analog amplifier). The results clearly highlight the limits of

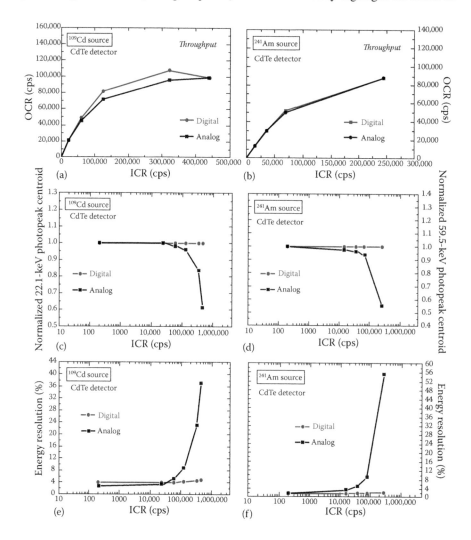

FIGURE 6.11 Performance comparison of the DPP system and the analog pulse processing electronics at various ICRs; both systems are characterized by similar throughput (OCR versus ICR curve). Throughput of both systems: (a) ^{109}Cd and (b) ^{241}Am sources. Normalized (c) 22.1- and (d) 59.5-keV photopeak centroids versus ICR. Energy resolution at (e) 22.1 keV and (f) 59.5 keV versus ICR.

the analog chain at high ICRs. The ^{109}Cd spectra measured with the analog chain are characterized by a peak centroid shift that is equal to 39% at 474 kcps (45% for ^{241}Am spectra at 244 kcps) and by high-resolution degradation. These distortions are due to baseline shifts and pileup effects that are more severe at high ICRs. On the contrary, the spectra acquired by the DPP chain are characterized by no peak position shifts and better energy resolution: 4.9% FWHM at 22.1 keV at 474 kcps (analog: 37%) and 2.3% FWHM at 59.5 keV at 244 kcps (analog: 55%). Figure 6.12 shows the high-rate spectra that were measured with both systems. To better highlight the potentialities of the DPP system or for high throughput or for optimum energy resolution, we performed measurements at various OCRs. Figure 6.13 shows the measured energy resolution versus throughput (i.e., OCR_{SLOW}/ICR) for some ICR conditions. The different throughput conditions are obtained by varying only

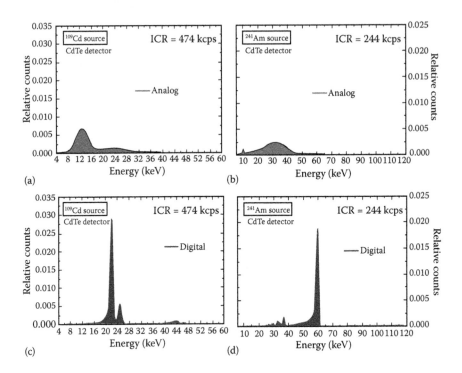

FIGURE 6.12 ^{109}Cd and ^{241}Am spectra measured with the CdTe detector by using the analog (shaping time constant value of 0.5 μs) and the DPP system (ST of 3.4 μs and 4.2 μs for ^{109}Cd and ^{241}Am sources, respectively). (a, b) The high peak shift and the high energy resolution degradation highlight the limits of the analog system in compensating the baseline shifts and pileup effects, which are more severe at high ICRs. (c, d) These results highlight the good high rate capability of the DPP system. The counts were normalized to the total number of the detected events. Energy calibration was performed at low ICR.

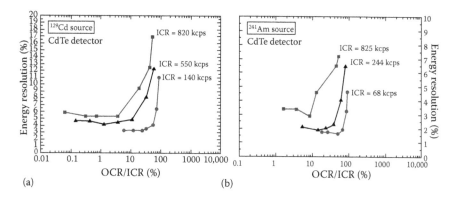

FIGURE 6.13   Energy resolution versus throughput (i.e., OCR$_{SLOW}$/ICR) for some ICR conditions. The energy spectra, (a) ^{109}Cd and (b) ^{241}Am sources, were measured with the CdTe detector and by using the DPP system. The different throughput conditions are obtained by varying the ST of the DPP system.

the ST of the DPP system. In this last analysis, the maximum rate range is fixed by the limits of the preamplifier (i.e., saturations occur at higher ICRs due to its energy rate limit). The energy spectra at low and high throughput are shown in Figure 6.14. We also used the PSD to minimize peak pileup events in the measured spectra. Figure 6.15 shows the W-target x-ray spectra that were measured at a moderate ICR of 136 kcps (30 kV; 2 mA) and at a very high ICR of 1.09 Mcps (30 kV; 16 mA). These measurements highlight the low degradation of energy resolution and of the energy calibration of the system even at very high ICRs.

6.4.2   MEASUREMENTS WITH A GE DETECTOR

^{137}Cs and ^{60}Co spectra were measured with the Ge detector at both low and high ICRs. Figure 6.16 shows the spectra that were measured with the DPP system at low ICRs. We used an ST of 25 and 30 μs for ^{137}Cs and ^{60}Co spectra, respectively. Energy resolutions (0.29% FWHM at 662 keV and 0.17% at 1333 keV) are similar to the values that were measured with the analog electronics (0.32% FWHM at 662 keV and 0.18% at 1333 keV, by using a shaping time constant value of 6 μs).

We also measured ^{137}Cs spectra at various ICRs (up to 55 kcps) with both analog and DPP systems, working at a similar throughput (inset of Figure 6.17a). To obtain a similar analog throughput, we used an ST of 25 μs (DPP) and a shaping time constant of 6 μs (analog). Figure 6.17 shows the energy resolution at various ICRs (Figure 6.17a) and energy resolution versus throughput at 55 kcps (Figure 6.17b). ^{137}Cs spectra measured with both systems at low and high throughput are shown in Figures 6.18 and 6.19.

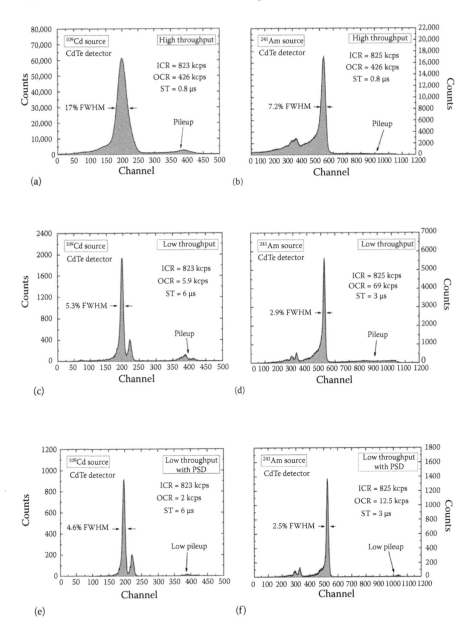

FIGURE 6.14 ^{109}Cd and ^{241}Am spectra measured with the CdTe detector. The spectra show the performance of the DPP system at high and low throughput. The different throughput conditions are obtained by varying the ST of the DPP system. The PSD technique was also used to reduce peak pileup (e) and (f).

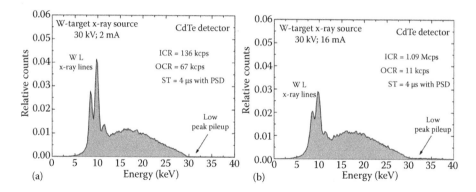

FIGURE 6.15 (a) Measured W-target x-ray spectrum at ICR of 136 kcps (30 kV; 2 mA). (b) Measured spectrum at very high ICR of 1.09 Mcps. The PSD technique was used to reduce peak pileup. The counts were normalized to the total number of the detected events. Energy calibration was performed at low ICR.

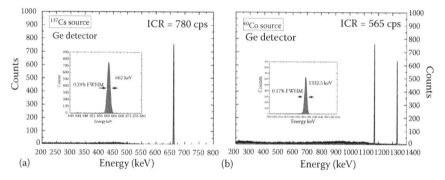

FIGURE 6.16 (a) ^{137}Cs and (b) ^{60}Co spectra measured with the Ge detector and the DPP system. Spectra exhibit energy resolution of 0.29% FWHM at 662 keV and 0.17% at 1333 keV.

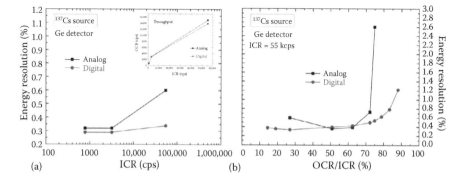

FIGURE 6.17 (a) Energy resolution at 662 keV versus ICR, measured with both the DPP and the analog systems working at a similar throughput. (b) Energy resolution versus throughput at 55 kcps. The different throughput conditions are obtained by varying the ST of the DPP system and the shaping time constant of the analog amplifier.

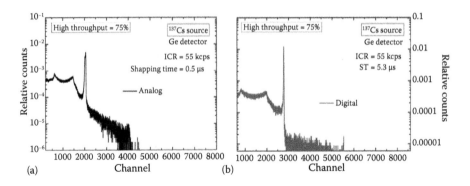

FIGURE 6.18 ^{137}Cs spectra measured at high throughput (75%) with (a) analog and (b) DPP systems. (a) Energy resolution of 2.6% FWHM at 662 keV and (b) energy resolution of 0.5% FWHM at 662 keV.

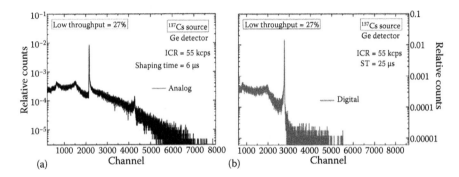

FIGURE 6.19 ^{137}Cs spectra measured at low throughput (27%) with (a) analog and (b) DPP systems. Energy resolution of 0.6% FWHM at 662 keV and (b) energy resolution of 0.3% FWHM at 662 keV.

REFERENCES

1. G. F. Knoll (ed.). 2000. *Radiation Detection and Measurement* (3rd ed.). New York: Wiley.
2. G. Bertuccio et al. 1996. Criteria of choice of the front-end transistor for low-noise preamplification of detector signals at sub-microsecond shaping times for x- and γ-ray spectroscopy. *Nucl. Instrum. Meth. A* **380** 301.
3. L. Abbene et al. 2007. X-ray spectroscopy and dosimetry with a portable CdTe device. *Nucl. Instrum. Meth. A* **571** 373.
4. U. Bottigli et al. 2006. Comparison of two portable solid state detectors with an improved collimation and alignment device for mammographic x-ray spectroscopy. *Med. Phys.* **33** 3469.
5. S. Stumbo et al. 2004. Direct analysis of molybdenum target generated x-ray spectra with a portable device. *Med. Phys.* **31** 2763.
6. A. La Manna et al. 2006. Portable CdTe detection system for mammographic x-ray spectroscopy. *Nuovo Cimento C* **29** 361.
7. M. Bolic, V. Drndarevic. 2002. Digital gamma-ray spectroscopy based on FPGA technology. *Nucl. Instrum. Meth. A* **482** 761.

8. J. M. Cardoso et al. 2004. CdZnTe spectra improvement through digital pulse amplitude correction using the linear sliding method. *Nucl. Instrum. Meth. A* **522** 487.
9. M. Arnold et al. 2006. TNT digital pulse processor. *IEEE Trans. Nucl. Sci.* **53** 723.
10. G. Gerardi et al. 2007. Digital filtering and analysis for a semiconductor x-ray detector data acquisition. *Nucl. Instrum. Meth. A* **571** 378.
11. T. Papp, J. A. Maxwell. 2010. A robust digital signal processor: Determining the true input rate. *Nucl. Instrum. Meth. A* **619** 89.
12. L. Abbene et al. 2010a. Performance of a digital CdTe x-ray spectrometer in low and high counting rate environment. *Nucl. Instrum. Meth. A* **621** 447.
13. L. Abbene et al. 2010b. High-rate x-ray spectroscopy in mammography with a CdTe detector: A digital pulse processing approach. *Med. Phys.* **37** 6147.
14. L. Abbene, G. Gerardi. 2011. Performance enhancements of compound semiconductor radiation detectors using digital pulse processing techniques. *Nucl. Instrum. Meth. A* **654** 340.
15. L. Abbene et al. 2012. Direct measurement of mammographic x-ray spectra with a digital CdTe detection system. *Sensors* **12** 8390.
16. M. Nakhostin, P. Veeramani. 2012. A new method for charge-loss correction of room-temperature semiconductor detectors using digital trapezoidal pulse shaping. *JINST* **7** P06006.
17. L. Abbene, G. Gerardi, F. Principato. 2013. Real time digital pulse processing for x-ray and gamma ray semiconductor detectors. *Nucl. Instrum. Meth. A* **730** 124.
18. L. Abbene et al. 2013. Energy resolution and throughput of a new real time digital pulse processing system for x-ray and gamma ray semiconductor detectors. *JINST* **8** P07019.
19. V. T. Jordanov et al. 1994. Digital techniques for real-time pulse shaping in radiation measurements. *Nucl. Instrum. Meth. A* **353** 261.
20. S. Del Sordo et al. 2004. Spectroscopic performances of 16 × 16 pixel CZT imaging hard-x-ray detectors. *Nuovo Cimento B* **119** 257.
21. S. Del Sordo et al. 2005. Characterization of a CZT focal plane small prototype for hard x-ray telescope. *IEEE Trans. Nucl. Sci.* **52** 3091.
22. S. Del Sordo et al. 2009. Progress in the development of CdTe and CdZnTe semiconductor radiation detectors for astrophysical and medical applications. *Sensors* **9** 3491.
23. S. Del Sordo et al. 2009. Recent trends in the development of CdTe and CdZnTe semiconductor detectors for astrophysical applications. *Proceedings of Science*, Otranto, Italy, 101520, October 13–17, 2009.
24. S. Del Sordo et al. 2009. Hard x-ray CZT detector development and testing on stratospheric balloon payloads. *ESA Special Publications* **671** 561.
25. L. Abbene et al. 2007. Spectroscopic response of a CdZnTe multiple electrode detector. *Nucl. Instrum. Meth. A* **583** 324.
26. L. Abbene et al. 2009. Hard x-ray response of pixellated CdZnTe detectors. *J. Appl. Phys.* **105** 124508.
27. N. Auricchio et al. 2011. Charge transport properties in CdZnTe detectors grown by the vertical Bridgman technique. *J. Appl. Phys.* **110** 124502.
28. F. Principato et al. 2012. Time-dependent current-voltage characteristics of Al/p-CdTe/Pt x-ray detectors. *J. Appl. Phys.* **112** 094506.
29. L. Abbene et al. 2013. Experimental results from Al/p-CdTe/Pt x-ray detectors. *Nucl. Instrum. Meth. A* **730** 135.
30. F. Principato et al. 2013. Polarization phenomena in Al/p-CdTe/Pt x-ray detectors. *Nucl. Instrum. Meth. A* **730** 141.
31. G.R. Gilmore. 2008. *Practical Gamma-Ray Spectrometry.* Chichester, West Sussex: John Willey & Sons, Inc.
32. Quality Assurance Data Sheet. GEM40P4-76, Ortec, United States.

7 Silicon Photomultipliers for High-Performance Scintillation Crystal Readout Applications

Carl Jackson, Kevin O'Neill, Liam Wall, and Brian McGarvey

CONTENTS

7.1 Introduction 142
7.2 A Short History of the SiPM 143
7.3 SiPM 143
    7.3.1 Geiger-Mode APD 143
    7.3.2 SiPM 145
    7.3.3 SiPM Output 146
    7.3.4 SiPM Operation in Detail 147
        7.3.4.1 Photon Capture 147
        7.3.4.2 Amplification 148
7.4 SiPM Sensor Parameters 150
    7.4.1 PDE 151
    7.4.2 Dark Count Rate 152
    7.4.3 Gain 154
    7.4.4 Crosstalk 156
    7.4.5 Afterpulsing 158
    7.4.6 Dynamic Range 159
7.5 SiPM Uniformity 160
    7.5.1 Breakdown Voltage 161
    7.5.2 Dark Current 161
    7.5.3 Optical Current 162
7.6 Packaging 165
    7.6.1 Poured Epoxy Packages 166
    7.6.2 Clear MLP 167
    7.6.3 TSV 170
    7.6.4 Summary Information 171
7.7 SiPM Reliability 172
    7.7.1 Issues of SiPM Reliability Assessment 172
    7.7.2 Reliability Assessment of SiPM 173

7.8   Scintillation Crystal Readout... 173
      7.8.1   Gamma-Ray Detectors 175
      7.8.2   Energy Resolution and Coincidence Timing Measurements........... 177
      7.8.3   SiPM Scintillation Crystal Performance 178
      7.8.4   CRT.. 179
      7.8.5   Energy Resolution.. 180
7.9   Summary 181
References.. 182

7.1   INTRODUCTION

Silicon photomultipliers (SiPMs) have emerged as the solid-state alternative to the vacuum tube that is known as the photomultiplier tube (PMT) in a number of scintillation detection applications ranging from medical imaging to radiation detection and isotope identification. The primary technical and commercial driver for the development of the SiPM has been medical imaging, specifically positron emission tomography (PET). In PET, there is a simultaneous requirement for higher-quality images and lower raw component sensor cost as compared to the PMT. The radiation detection market requires higher-performance sensors for better isotope identification, increased physical robustness for handheld and field operations, low operating voltage compared to the high voltage of the PMT, and reduced sensor cost for high-volume homeland security applications. A typical generic application example of an SiPM detecting the light emission from a scintillating crystal is shown in Figure 7.1. This figure shows the scintillation crystal that is coupled to an SiPM sensor, which is directly connected to a printed circuit board (PCB). The signal from the SiPM is typically processed via an amplifier and/or a shaper followed by data acquisition. The scintillation crystal is capable of converting the incident gamma rays to visible photons, and the SiPM sensor has the sensitivity to convert the visible photons to an electrical charge, which can be amplified, shaped, and read out by the electronics. The SiPM is typically mounted on a PCB, and due to its compact nature, it is possible to place the readout electronics on the same PCB.

      The SiPM can be used to replace the positive-intrinsic-negative (PIN) photodiode, the avalanche photodiode (APD), or the PMT that is currently used in these systems.

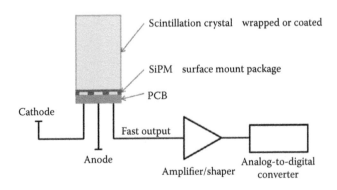

FIGURE 7.1   Simplified diagram of a detector and the readout block diagram for scintillation crystal readout.

For all applications, the main benefits of SiPM technology include high optical sensitivity, low operating voltage, unsurpassed uniformity across the sensor, uniformity from sensor to sensor, small form factor, and low cost. Specifically, the sensors discussed in this work will be complementary metal–oxide semiconductor (CMOS)-manufactured SiPM sensors, which have been developed to meet the performance and cost requirements of medical imaging PET and radiation detection markets.

7.2   A SHORT HISTORY OF THE SiPM

The SiPM sensor, now in mass production, has been developed through the knowledge of the Geiger-mode operation of photodiodes, which can be traced back to the original work by Haitz[1-4] and Oldham et al.[5] The Geiger-mode photodiode was used as a tool to investigate diode breakdown properties and microplasmas in diodes and helped to develop the diode impact ionization and avalanche breakdown models. The work of McIntyre[6,7] developed equations allowing for a full understanding of the device physics in a practical and workable model of breakdown and above-breakdown Geiger-mode operation. Further, published work on the Geiger-mode technology during the 1980s and 1990s centered on using the Geiger-mode photodiode for single-photon counting applications.[8] This work was largely focused on individual photodiode sensors with external quenching circuitry.[9,10] In the late 1990s and early 2000s, the SiPM concept emerged from Russia through work that was published by Saveliev,[13] Bisello,[14] and Golovin.[15] However, the development of the SiPM concept can be traced to papers that were published in the 1970s by Krachenko et al.[11,12] in which arrays of Geiger-mode avalanche sensors, passively quenched, are used for low-light sensing. At the time of publication in 1978, the authors noted that a limitation of p-type to n-type (PN) photodiodes as an avalanche-based sensor is the presence of microplasmas in the photodiode depletion region, which limits the useful size of the photodiode's active area. This limitation was removed in the late 1990s and 2000s as Geiger-mode photodiodes were developed utilizing higher-quality start material and processing conditions, and architectures were developed that reduced the presence of defects in the depletion region of the photodiode. This advance allowed large Geiger-mode photodiode arrays to be fabricated. These developments allowed the SiPM to be manufactured in CMOS foundries, creating a cost-competitive alternative to the PMT.

7.3   SiPM

In this section, the SiPM will be described in more detail, from the basic theory of operation to the major parameters that govern sensor performance. The basic theory of the SiPM will be developed from the theory of Geiger-mode operation of single microcells to arrays of microcells that form the model for the SiPM.

7.3.1   GEIGER-MODE APD

This basic building block of an SiPM is a photodiode that is sensitive to single photons, possesses high internal gain, and self-regulates its output current. A circuit schematic of the basic structure that accomplishes this is shown in Figure 7.2a. The photodiode

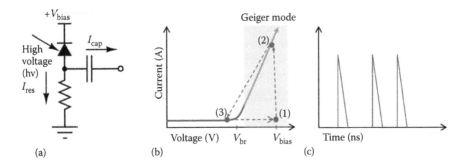

FIGURE 7.2 Basic schematic and operation of a Geiger-mode APD. (a) A single microcell of a SiPM including a single photon avalanche diode (SPAD), quench resistor and fast output capacitor. (b) The current voltage curve of the circuit from (a). (c) The output voltage pulses from (a).

in Figure 7.2a is designed to operate in Geiger mode so that it can detect single photons. The resistor in series with the photodiode provides the means of limiting the current flow and resetting the photodiode after each photon is detected. The Geiger-mode operation can be explained by reviewing the current-versus-voltage characteristic that is shown in Figure 7.2b. This figure shows the ideal reverse-bias characteristics for a photodiode and overlays the mode of operation, known as Geiger mode, which allows for single-photon detection. A photodiode has a breakdown voltage (V_{br}) at which point the current flow in the diode significantly increases. For operation in Geiger mode, a voltage in excess of V_{br} is applied, as shown in Figure 7.2b (V_{bias}). The difference in voltage between V_{bias} and V_{br} is termed the overvoltage, and typical values are between 1 and 5 V. Initially, upon application of V_{bias}, no current flows in the photosensitive region of the photodiode. There is always a small parasitic current, shown in Figure 7.2b, flowing in the photodiode, which to the first order can be neglected. Upon the arrival of a photon and conversion of the photon into an electron–hole pair, the charge carriers undergo impact ionization, leading to avalanche multiplication. During the avalanche process, the photodiode current changes from a no-current state, labeled as (1) in Figure 7.2b, to a high-current state, labeled as (2). The current generated in the photodiode follows two paths in Figure 7.2a, which are shown as I_{res} and I_{cap}. Once I_{res} begins to flow through the photodiode, voltage is dropped across the series resistor. The voltage drop causes the photodiode voltage and electric field to decrease, and the operating point shown in Figure 7.2b moves from (2) to (3). As a current is no longer flowing across the resistor, the voltage bias across the photodiode increases to the original V_{bias}, and the photodiode returns to a state that can detect the next photon. The output from this type of sensor is shown in Figure 7.2c and can be used to explain both I_{res} and I_{cap}.* The output current is characterized by a sharp and fast rise time at the onset of the incident photon, relating to the transition from (1) to (2) and slower recovery time as the photodiode transitions from (3) back to (1). As will be shown in Section 7.3.3, the addition of a capacitor to the photodiode and resistor provides a simple means of accessing the fast-switching behavior inside the SiPM.

---

* I_{cap} is available on products from SensL to allow for a low-capacitance, fast output. Other SiPMs do not contain I_{cap}.

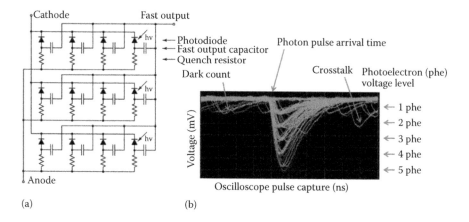

(a)   (b)

FIGURE 7.3 (a) SiPM schematic. (b) Characteristic output response of an SiPM as viewed on an oscilloscope with the persistence setting engaged.

The output signal from the Geiger-mode APD consists of nearly identical current pulses in time. Each current pulse relates directly to the detection of a photon or thermal noise event in the sensor. The thermal noise, known as the dark count rate, has the same properties as a single photon and will be shown to be a key parameter for the understanding of the SiPM operation.

7.3.2 SiPM

The Geiger-mode APD provides single-photon sensitivity and outputs a current pulse that signals the arrival time of the incident photon. However, the Geiger-mode APD is a binary sensor and does not have the ability to sense multiple-photon events since the output can be considered identical regardless of photon number. Additionally, the size of the photodiode is limited by two effects: (1) the characteristic dark count noise increases with active area, and (2) the recovery time (during which the sensor is insensitive to further incident photons) increases with increasing capacitance of the diode, which in turn increases with the active area. Therefore, only relatively small Geiger-mode APDs are used, typically tens to hundreds of micrometers in diameter. To allow for both a large-area sensor and to detect multiple photons, the SiPM was created. The SiPM has a structure as shown in Figure 7.3a. A key feature of the SiPM architecture is the large array of parallel-connected Geiger-mode APDs with quenching elements and output capacitors* that are connected together. Each photodiode, resistor, and capacitor element are collectively called a microcell. The schematic in Figure 7.3a shows a simplified SiPM that is demonstrated as a 4 × 3 array of microcells. A production SiPM would have from hundreds to thousands of microcells per square millimeter. The SiPM is biased with a positive voltage on the cathode with respect to the anode. The SiPM output can be measured as the current flow from the cathode to the anode or from the fast output. A representation of the

---

* Fast-output capacitors are only included on SiPMs that are provided by SensL.

output of an SiPM in response to a low-level light pulse is shown in Figure 7.3b. This image was captured on an oscilloscope with the persistence setting set to allow for displaying multiple detected events. To create the voltage signal, the output of the SiPM was connected to an external resistive load.

Figure 7.3b shows several key features of the SiPM output such as (1) the ability to detect multiple-photon events, (2) the dark count signal, which is at the same level as the single photoelectron signal, and (3) crosstalk, which is an unwanted feature that is caused by cotriggering inside the SiPM. The detection of multiple photons is shown as the clearly discernable voltage levels that are labeled as 1–5 phe. This represents the detection of between one and five single photons. The photon number detection is a desired characteristic, whereas the dark count and crosstalk are the negative features of the SiPM operation. Reviewing the architecture in Figure 7.3a, it is clear that in an SiPM, there is the optically active photodiode as well as optically inactive structures such as the resistor and the capacitor. The necessity to provide physical separation for the individual microcell photodiode elements required for independent Geiger-mode operation means that there is a dead space between photo-diodes that limits the total active area. The percentage of active area in an SiPM is termed the fill factor and must be considered with the number of microcells that are found in each SiPM. Higher fill factor increases photon detection efficiency (PDE) at the expense of dynamic range. We will cover the different aspects of this issue in more detail in Section 7.4.1.

7.3.3   SiPM Output

The SiPM, as shown in Figure 7.3a, has three outputs. The connection for biasing the photodiode is made through the cathode and anode terminals. In this exam-ple, the anode is directly connected to the quenching resistor. However, this is still termed the anode since it assists the user in the application of the correct bias voltage for reverse bias operation. The additional fast output is shown as a capacitively cou-pled output, which is directly connected to the microcell at the intersection between the anode and the quench resistor. The output signal is shown in Figure 7.4 for a fast-pulsed laser with a pulse width of less than 100 ps. The anode–cathode output is characterized by a fast onset time that is followed by a recovery of the output signal, which is governed by the internal impedance of the SiPM and the external circuitry that is directly connected to the sensor. The recovery time increases with the photodiode area and the quench resistor and parasitic impedance. The fastest recovery times are achieved with the smallest microcell size but at the expense of the fill factor. The fast output signal is characterized as a short fast pulse, which is the derivative of the switching behavior of the photodiode during Geiger-mode operation that is detailed in Figure 7.2b (1 and 2). The decay time, which is a feature of the anode–cathode signal, is not present; instead, this is replaced by a small undershoot to the baseline voltage as the microcell recovers. The fast output represents the signal accurately, and since the recovery time is removed from the output signal, it is easier to identify additional incident photons within the recovery period. It will be shown that the scintillation crystal readout can be achieved with excellent energy and coin-cidence timing resolution using only the signal from the fast output.

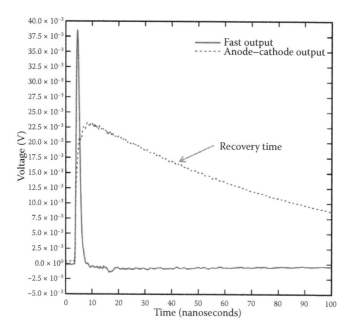

FIGURE 7.4 Pulse output from 3 × 3-mm² SiPM with 35-μm microcells. Fast output and anode–cathode output are shown superimposed on the same graph.

7.3.4 SiPM Operation in Detail

Sections 3.1 through 3.3 describe the basic operation of the SiPM. In this section, we will review the theory of operation in additional detail. As has been shown, the SiPM provides the ability to detect individual photons and provides an output current, which is proportional to the number of incident photons. This is a key factor for the readout of the light emission from a scintillation crystal, as will be discussed in Section 7.8.1. For the SiPM to be able to detect the photons that are emitted from scintillation crystals and output current pulses, as shown in Figures 7.3b and 7.4, the following processes must occur:

- *Photon capture*—the photon must be absorbed and converted into an electron–hole pair.
- *Amplification*—the electron–hole pair must be converted into a measurable current through the internal gain mechanism of the SiPM.

7.3.4.1 Photon Capture

Photons entering the SiPM are absorbed according to the Beer–Lambert equation for photon absorption, which is given as

$$I(z, \lambda) = I_0 e^{-\alpha(\lambda)z}, \tag{7.1}$$

where I_0 is the incident photon flux (typically photons per second), $\alpha(\lambda)$ is the attenuation coefficient of silicon at the wavelength of interest, and z is the distance into the SiPM. This equation determines the ability of an SiPM to detect photons of a given wavelength. Using the scintillator peak emission values given in Figure 7.26 and combining this with the attenuation coefficients for crystalline silicon,[19] it is possible to calculate the percentage of photons that are absorbed as a function of distance into the SiPM. This calculation is shown in Figure 7.5 for various peak intensities of scintillators that are relevant to this work. LaBr3 emits photons with a peak wavelength at 380 nm, and Figure 7.5 shows that almost 100% of them are absorbed at a distance of 0.2 μm into the SiPM. Conversely, at 0.2 μm, only 10% of the photons from CsI(Tl), which has an emission peak at 550 nm, are absorbed.

A longer emission wavelength requires increased silicon depth to allow the efficient capture of the incident photons. However, the scintillators described here all emit photons in a range that can be sensed by SiPM devices that possess shallow junctions on the order of 0.2–1 μm, which are the subject of this work. These SiPMs rely on the planar, shallow structures, which CMOS processing is ideally suited to fabricate.

7.3.4.2 Amplification

Once a photon has been absorbed by the SiPM, the next step is the amplification of the generated electrons and holes. This amplification is governed by impact ionization of electrons and holes in the depletion region of the photodiode. The depletion region of a P-on-N photodiode, shown in Figure 7.6a, is the region of a high electric

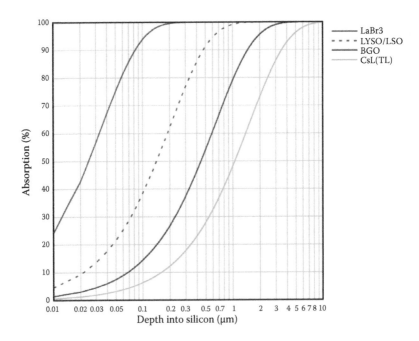

FIGURE 7.5 Photon absorption in SiPM sensors for various crystal peak emission wavelengths.

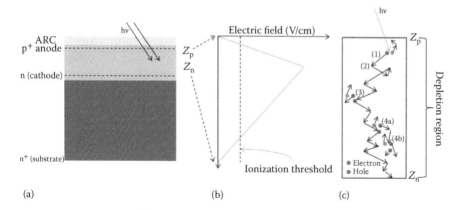

(a)   (b)   (c)

FIGURE 7.6 (a) Simplified cross section of a P-on-N Geiger-mode APD structure. (b) Electric field profile across the photodiode depletion region. (c) A pictorial representation of avalanche breakdown in a photodiode across the depletion region.

field that is established where the N- and P-type doped silicon layers come together. This dopant structure is fully compliant with standard CMOS processing, which is used for production of high-quality SiPMs in volume. The anode of the photodiode for a P-on-N structure is located close to the surface beneath an antireflection coating, which has been designed to maximize photon transmission into the silicon. The cathode of the photodiode is below the anode and electrically connected to a low-resistance substrate of the same dopant type. To put the physical sizes in perspective, a typical silicon wafer used to manufacture an SiPM is on the order of 750 μm in thickness, whereas the depletion region with a breakdown voltage of 24.5 V is on the order of 1 μm in thickness. Most of the active volume of the SiPMs designed for radiation detection is located close to the surface, and the substrate typically provides a low ohmic contact to the cathode and the necessary structural support.

The SiPM avalanche process is governed by the electric field across the depletion region of the photodiode, as shown in Figure 7.6b. With knowledge of the cathode and anode dopant levels, Poisson's equation can be used for the calculation of the electric field across the junction. For the discussion here, it is sufficient to know that the depletion region electric field is increased through the application of a reverse-bias voltage. When a reverse bias is applied to the SiPM in excess of the breakdown voltage, the electric field increases beyond the ionization threshold. At this point, charge carriers gain sufficient energy to create further electron–hole pairs through collisions, a process called impact ionization. If the electric field is sufficiently high, then avalanche breakdown will occur. Avalanche breakdown is pictorially represented in Figure 7.6c as a succession of charge carrier movements that are labeled 1 through 4. As the charge carrier passes through the depletion region, it is accelerated by the electric field. The electric field accelerates electrons toward the cathode and holes toward the anode. The acceleration increases the carrier kinetic energy, but carriers cannot pass through the depletion region without interacting with the underlying silicon structure. There are two main types of interactions that can occur: (1) phonon generation or (2) impact ionization. Phonon generation is shown as 1 and 2 in Figure 7.6c and results in a scattering event that reduces the kinetic energy of the charge carrier. Ionization events are shown as 3, 4a, and 4b in Figure 7.6c. The ionization causes the generation of additional electron–hole pairs, which can themselves generate ionization events, as shown in 4b. With a sufficiently high electric field, the impact ionization process can, in theory, continue indefinitely. However, with the inclusion of the quench resistor in the microcell, as shown in Figure 7.2a, the electric charge at each breakdown event, and therefore the current, is limited to only the capacitance of the microcell's photodiode. The avalanche process provides the output current and the characteristic breakdown voltage of the SiPM.

7.4 SiPM SENSOR PARAMETERS

It is important to discuss the major sensor parameters that impact the performance characteristics of the SiPM. The following are considered the basic SiPM sensor parameters:

- *Breakdown voltage*—the voltage at which the SiPM current increases rapidly with increasing reverse bias
- *Photon detection efficiency*—or PDE, the absolute photon detection efficiency of the SiPM including all geometrical effects
- *Dark count*—the number of false single-pulse height signals that are measured per second
- *Gain*—the internal gain of the SiPM in response to a single photon
- *Crosstalk*—output signal that is caused by one microcell randomly firing one or more additional microcells in the array
- *Afterpulsing*—secondary avalanche events, which are triggered by stored charge during a previous event
- *Dynamic range*—the ability of an SiPM to detect incoming photons without saturation

To aid in the understanding of the sensor parameters discussed, Table 7.1 is used. This details the various SiPM types that are available, including active area, microcell size, the number of microcells, and the fill factor. It is important to note that the microcell size used in this table is the size of the active area of the microcell and

TABLE 7.1
Physical Parameters for the SiPM Discussed in This Work

Active Area Dimensions	SiPM Type	Microcell Size	Number of Microcells	Fill Factor	Recovery Time (1/e)
1×1 mm^2	10035	35 μm	576	64%	82 ns
3×3 mm^2	30020	20 μm	10,998	48%	41 ns
	30035	35 μm	4774	64%	82 ns
	30050	50 μm	2668	72%	159 ns
6×6 mm^2	60035	35 μm	18,980	64%	96 ns

not the pitch of the microcells in the SiPM. These parameters are from the current production of SiPM by SensL.[22]

Each of these parameters will impact the radiation detection performance when the SiPM is coupled to a scintillating crystal. The individual parameters are discussed in the next section including the information of their accurate measurement.

7.4.1 PDE

The PDE is a key parameter for SiPM operation. This parameter determines the percentage of photons that will be detected by an SiPM. The PDE of an SiPM can be defined by the following equation:

$$PDE = QE(\lambda) * AIP(V) * FF, \tag{7.2}$$

where $QE(\lambda)$ is the internal quantum efficiency of the SiPM, AIP is the avalanche initiation probability, and FF is the fill factor of the SiPM. The QE is wavelength dependent and includes any losses at the optical interface of the sensor. The AIP is set by the doping levels and physics of the depletion region of the photodiode and increases with voltage. The FF is a geometrical parameter that is governed by the ratio of active area to nonactive area in the SiPM microcell.

To measure the PDE, an accurate measurement must be made that utilizes photon statistics. In photon statistics inherent in SiPM sensors, there is a specific relationship between the ratio of events that trigger zero photons, N_0, and the total number of events that include any total number of photons, N_{tot}. This relationship is determined through the chance of observing a zero-photon event in a Poisson probability distribution with a mean number of photons, n_{ph}, and can be written as

$$\frac{N_0}{N_{tot}} = e^{-n_{ph}}.$$

To correctly take account of background pulses, a measurement in the dark must be obtained. The parameters N_0^{dark} and N_{tot}^{dark} must be measured, and then the mean number of photons n_{ph} arriving at the SiPM can be calculated using

$$n_{ph} = -\ln\left(\frac{N_0}{N_{tot}}\right) + \ln\left(\frac{N_0^{dark}}{N_{tot}^{dark}}\right) \tag{7.3}$$

The PDE can then be calculated by dividing the mean rate of detected photons by the recorded rate of arriving photons, given by the optical power that is recorded by a calibrated photodiode P and the period T of the light pulses:

$$PDE = \frac{n_{ph}R/T}{P/\frac{\lambda}{hc}} \tag{7.4}$$

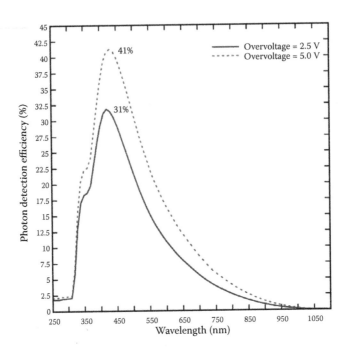

FIGURE 7.7   PDE versus wavelength for 30035 SiPM at overvoltages of 2.5 and 5.0 V.

Here, R is the geometrical optical power ratio between the light falling on the SPM and the calibrated photodiode. An example of this measurement technique that is extensively referenced is the work of Eckert.[1] This technique produces the true SiPM PDE that does not contain the effects of crosstalk or afterpulsing.

Figure 7.7 shows the PDE for a 30035-type SiPM across the wavelength range of 250–1100 nm. Peak emission is shown at 420 nm with a maximum peak detection efficiency of 41% that is achieved at 5.0-V overvoltage.

The PDE varies with the microcell size according to the fill factor for the different types of SiPM. The PDE does not vary with the active area dimension. Therefore, the fill factor values listed in Table 7.1 can be used to scale the PDE that is shown in Figure 7.7 to different sensors. To scale the PDE to different sensor types, first, divide the PDE of the known SiPM by the fill factor of the known SiPM, and then multiply by the fill factor of the unknown SiPM. As an example, to scale the 5.0 overvoltage PDE at 420 nm from the 30035 to the 30050 SiPM, we will use the following equation:

$$41\%/0.64 * 0.72 = 46\%.$$

7.4.2 DARK COUNT RATE

The dark count rate is defined as the pulse rate that is measured in the dark, with a leading-edge trigger at 0.5 times the single photoelectron amplitude. The single photo-electron amplitude is determined by observing when the count rate first decreases as the leading-edge threshold is increased. By increasing the threshold until the dark count rate changes, it is possible to determine the maximum pulse height of the single photoelectron counts. A measurement of the dark count rate is not impacted by crosstalk but can be influenced by afterpulsing. In Section 7.4.5, afterpulsing will be shown to be very low for the SiPM sensors that are discussed here and is not a significant factor in the measured dark count rate. The dark count rate for 10035, 30035, and 60035 SiPM is shown versus overvoltage in Figure 7.8. This demonstrates a very well-controlled dark count rate versus overvoltage that is suitable for mass production applications that require SiPM sensor output stability over a wide range of overvoltage conditions and across a wide range of temperatures.

For scintillators with a short decay time, the dark count rate can be largely ignored. For example, in Table 7.4, it is shown that LaBr3 has a decay time constant of 16 ns, and, typically, pulses are integrated on the order of 100 ns to achieve good energy resolution and full signal collection. For a 100-ns signal integration time using a 60035 sensor, and using the dark count values, as shown in Figure 7.8, the average number of dark counts integrated can be calculated as $5 \times 10^7 * 100 \times 10^{-9} = 5$. The dark count rate is clearly not a dominant factor for a LaBr3, which emits ~63 photons/keV and therefore would emit >40k visible photons in response to a single 662-keV gamma ray. For scintillators with longer characteristic decay times such as CsI(Tl), combined with the desire to detect low-energy gamma rays in tens of kiloelectronvolts, the dark count rate becomes a more important factor. The dark count rate can degrade the signal/noise ratio. Advances in SiPM production are leading to improved dark count rates. An example of the next-generation dark count

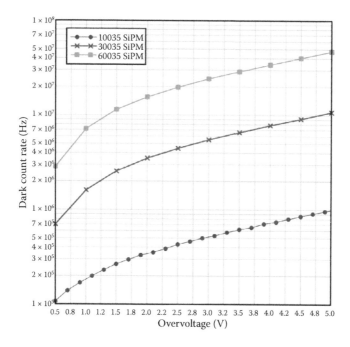

FIGURE 7.8 Measurement of dark count rates for 10035, 30035, and 60035 SiPM.

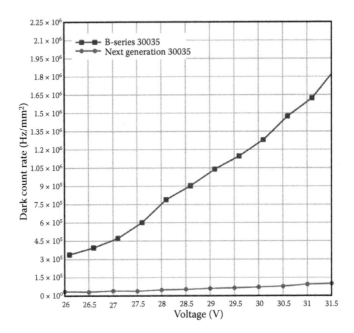

FIGURE 7.9 Dark count rate of next-generation SiPM. Dark count rates are based on 35-μm-based microcell sensors representing dark count rates per square millimeter for 10035, 30035, and 60035 sensors.

rates for SiPM is shown in Figure 7.9, compared to the existing family of SiPM. This shows the achievement of a dark count rate of sub 100 kHz/mm².

7.4.3 GAIN

The gain of an SiPM is defined as the mean number of charge carriers that are generated when a single charge initiates an avalanche process in the depletion region of the sensor. This is a unitless parameter and determines the amount of charge flowing through the sensor during detection. The gain can be derived theoretically from the following equation:

$$G = C_{microcell} * \frac{V_{bias} - V_{br}}{q}, \tag{7.5}$$

where $C_{microcell}$ is the capacitance of a single microcell, V_{bias} is the operating voltage, V_{br} is the breakdown voltage, and q is the electronic charge.

Knowledge of the gain of the SiPM aids in conversion between the dark count rate and the measured dark current in the SiPM by the following equation:

$$I_{dark} = DCR \times G \times q \times (1 + CT) \times (1 + AP), \tag{7.6}$$

where I_{SiPM} is the current in the SiPM, DCR is the dark count rate in Hz, G is the gain, CT is the crosstalk probability, and AP is the afterpulse probability. It is possible to approximate the SiPM by ignoring the CT and AP terms if they are deemed sufficiently small. This is appropriate for first-order calculations, but crosstalk and afterpulsing should always be considered for the most accurate estimations.

To measure the gain of an SiPM requires knowledge of the current flowing through the sensor that is measured while in saturation using a pulsed light source. A standard pulsed light-emitting diode (LED) of 470 nm can be used for the measurement. The basic concept of the measurement is as follows: if all the microcells of an SiPM are illuminated simultaneously and with a known repetition rate, it is possible to infer the gain of the sensor by measuring the DC current flowing through the standard anode–cathode. Once this is obtained, it is necessary to subtract the dark current I_{dark} from the photocurrent I_{photo}. The gain is then

$$G = \frac{I_{photo} - I_{dark}}{q * f_{laser} * N},$$ (7.7)

where q is the electron charge, f_{laser} is the laser pulse frequency, and N is the number of microcells. The gain measurement completed in this way does not contain afterpulsing or crosstalk and is a true representation of the gain of an SiPM. A plot of the gain for the different SiPM microcell types is shown in Figure 7.10. No difference

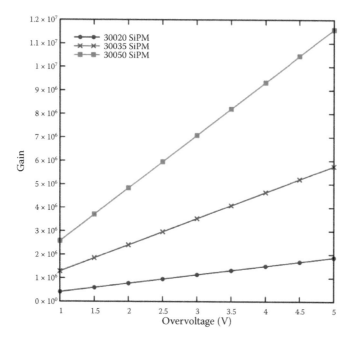

FIGURE 7.10 Gain measurement for standard anode–cathode output for B-Series 30020, 30035, and 30050 SiPM.

was noted between the samples of different active area types that utilize the same microcell structure.

The gain values for the fast-output terminals were designed to be between 2% and 4% of the standard anode–cathode gain and set so that the gain was not dependent on the microcell size. A measured gain for the fast output was therefore found to be 4.3×10^4 regardless of microcell type at 2.5-V overvoltage. The gain measurement technique was also used to accurately determine the temperature coefficient of the breakdown voltage. The point of the zero-crossing gain versus voltage curve was measured at a range of temperatures. The breakdown voltage was found to vary by 21.5 mV/°C over standard operating temperature ranges.

7.4.4 CROSSTALK

A crosstalk event can be seen in the output signal of an SiPM, as shown in Figure 7.3b. Crosstalk is defined as the ratio of the pulse rate that is measured at 0.5 and 1.5 times the single photoelectron amplitude. Crosstalk is an undesirable phenomenon in that a single-firing microcell fires one or more neighboring microcells almost instantaneously. Crosstalk pulses can therefore appear as multiple output signals in response to a single photon and is therefore a form of multiplication noise impacting the excess noise factor (ENF). Increasing bias voltage increases crosstalk. Crosstalk can be caused by electrical charge leakage or optically generated due to the emission of light in SiPM sensors.

To measure the crosstalk in SiPM sensors, the rate of dark counts as a function of an increasing leading-edge trigger threshold is measured and repeated at various bias voltages. An example of the output from this type of measurement is shown in Figure 7.11. The y-axis indicates the frequency of the dark count that is measured for the input trigger threshold on the x-axis.

As shown in Figure 7.11, this results in a staircase dependence of rate-versus-trigger threshold and progresses to lower rates as the trigger threshold increases. The 1.5 (height) and 0.5 (height) photoelectron rates can then be measured, and the ratio is computed for each overvoltage. Using this procedure, the crosstalk measured for 30020, 30035, and 30050 SiPM is shown in Figure 7.12.

Crosstalk was not seen to depend on the size of the individual die and has been found to only depend on the size of the microcells. To investigate this phenomenon, the crosstalk on 10035 and 60035 SiPM was measured. The crosstalk on these SiPMs was within measurement accuracy for the 30035 SiPM and can be considered identical regardless of die size. For the 30020 sensors, the crosstalk rates were found to be extremely low and difficult to measure without long measurement cycle times. The measurements below 2-V overvoltage were in fact terminated due to the length of time taking for the measurement. These 10020 SiPMs exhibit a crosstalk of sub 6% at the 5.0-V level. In all cases, the crosstalk at typical operating overvoltages for the SiPM reported in this work was found to be extremely low. It is expected that for most applications, the 30035 SiPM will provide the best combination of low crosstalk and high PDE.

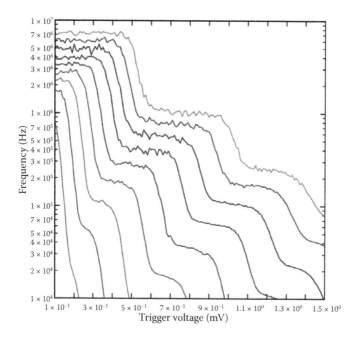

FIGURE 7.11   Crosstalk threshold plot for 30035 SiPM. Data are taken from 1- to 5-V overvoltage.

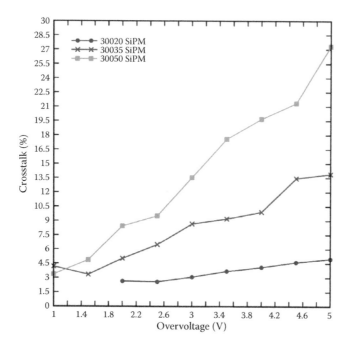

FIGURE 7.12   Measurement of crosstalk for 30020, 30035, and 30050 SiPM.

7.4.5 AFTERPULSING

Afterpulsing is the phenomenon of an SiPM microcell randomly discharging with higher probability shortly after a previous discharge than at the expected thermal generation rate. Afterpulse events that occur after the recovery time cannot be distinguished from genuine photon-induced events and therefore deteriorate the photon-counting resolution of an SiPM. The typical time scales for this phenomenon are tens of nanoseconds, comparable with the microcell recovery time. As a result, many afterpulse events are of partial discharge that is compared with the single photoelectron discharge of a microcell, due to partial recharge of the microcell.

Afterpulsing is an undesirable effect because it increases both the variance of the single cell charge and crosstalk in the SiPM and reduces dynamic range. The degree of afterpulsing depends on the bias voltage, temperature, the design of the deep doping implant, and the recovery time.

Afterpulse probability measurements are made by binning the time between consecutive pulses and extracting via the formula

$$P_{ap} = \frac{\int_0^\infty \xi * n_{ap} d\Delta t}{\int_0^\infty \xi * (n_{ap} + n_{tp}) d\Delta t}, \tag{7.8}$$

where

$$n_{tp}(\Delta t) = \frac{N_{tp}}{\tau_{tp}} * e^{-\frac{\Delta t}{\tau_{tp}}}, \tag{7.9}$$

$$n_{ap}(\Delta t) = \frac{N_{apf}}{\tau_{apf}} * e^{-\frac{\Delta t}{\tau_{apf}}} + \frac{N_{aps}}{\tau_{aps}} * e^{-\frac{\Delta t}{\tau_{aps}}}. \tag{7.10}$$

Here, n_{tp} is the thermal rate probability density, n_{ap} is the afterpulse probability rate, consisting of fast and (optionally) slow components, and ξ is the charge fraction of a second pulse time Δt after a given pulse. Afterpulsing was found to be dependent only on the microcell geometry and not on the SiPM die size. In Figure 7.13, the afterpulse rates measured for 30020, 30035, and 30050 SiPM are shown.

The results obtained for 10035 and 60035 SiPM were within measurement error to the 30035 SiPM. This demonstrates the low level of afterpulsing that is present in the SiPM, and for the 30020 and 30035, afterpulse rates are less than or equal to 5% up to 5.0-V overvoltage. The afterpulsing measured on the 30050 was found to be similar to the 30035 until 4.5 V. After 4.5 V, the afterpulsing of the 30050 increased more rapidly than the 30035. It was believed that the high gain of the 30050, as shown in Figure 7.10, was the cause for the increase in afterpulsing at overvoltages in excess of 4.5 V. The low level of afterpulsing in the SiPM presented demonstrates the

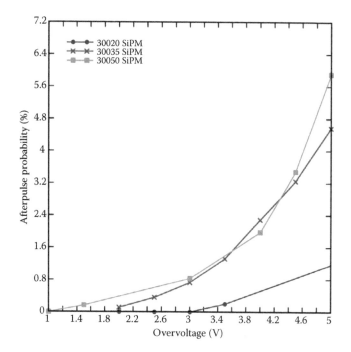

FIGURE 7.13 Afterpulse measurements on 30020, 30035, and 30050 SiPM.

high quality of the start material and the processing steps that are used to fabricate these SiPM sensors.

7.4.6 Dynamic Range

The dynamic range of the SiPM is determined by the number of microcells over the active area that is available to detect incoming photons. For fast-scintillation-decay-time crystals, such as LaBr3 and lutetium yttrium orthosilicate (LYSO) (Table 7.4), the decay time is shorter than the decay time of a typical SiPM (Table 7.1), and to the first order, the crystal emission can be considered as a pulse source. For longer-decay-time crystals such as CsI(Tl), this becomes more complicated as microcells can recover and respond multiple times during a scintillation emission. For simplicity, we will only consider the pulse dynamic range here. To the first order, the dynamic range of the SiPM can be considered simply by the following equation:

$$N_{\text{fired}}(M,\lambda) = M\left(1 - \exp^{\left(-\frac{\text{PDE}(\lambda)N_{\text{ph}}}{M}\right)}\right), \tag{7.11}$$

where M is the number of microcells per die, PDE is the photon detection efficiency at a specified lambda, and N_{ph} is the number of incident photons. A characteristic curve can be plotted for this equation using basic information that is derived from the datasheet of SiPM products. For Figure 7.14, the dynamic range for pulsed light is

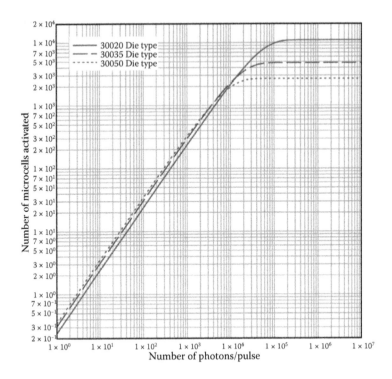

FIGURE 7.14 Dynamic range calculations for 30020 (10,998 microcells), 30035 (4774 microcells), and 30050 (2668 microcells) die types.

shown from the single photon level up to 10^7 photons. For each die type, the parameters of M and PDE have been obtained from the datasheet for the B-Series SiPM.[22]

Taking the photon yield of a scintillator such as LYSO from Figure 7.26 as 32 photons/keV, and assuming an incident gamma-ray photon of 511 keV, approximately 16,352 visible photons should be emitted. Assuming a 50% loss from the light collection and transfer to the SiPM and a PDE of 41% at the peak wavelength of 420 nm from Figure 7.7, there should theoretically be ~3352 photons that are incident on the sensor. From Figure 7.14, it can be seen that with this number of incident photons on the sensor, it would generate a signal that is equivalent to ~700 microcells firing. Additionally, we can see that this signal level is at the top end of the linear range of the sensor. For this reason, the 30035 sensor type with 35-μm microcells works well with LYSO crystals, though careful analysis of the linearity should be performed to produce the most accurate energy resolution measurements.

7.5 SiPM UNIFORMITY

The uniformity of the sensor operating parameters is important for the use of SiPM sensors in high-volume applications. The sensor uniformity minimizes variation between sensors and reduces or eliminates voltage adjustment between the sensors that are used in a system. To demonstrate uniformity, this section will review the end-of-line test data that are obtained during the manufacturing of SiPM. The sensors described in this section are from SensL's B-Series sensors. These sensors are manufactured on 200-mm wafers in a CMOS foundry, typically in lots of 25 wafers. Following the completion of the fabrication process, each wafer is comprehensively tested, first at wafer acceptance test (WAT) using a process control monitor (PCM) and then at component probe (CP) using a product test. The WAT verifies that the wafer was processed correctly and that key technological parameters such as breakdown voltage and diffused, and deposited film resistances are on target and within the specified test limits. These test limits were established based on technology target values, material, and process characteristics such as film thickness and resistivities and from the data that were collected during the technology development phase from skew or split lots. The measurement data from the WAT are monitored wafer to wafer and lot to lot, and trends were analyzed to verify process capability and stability. This process is a standard part of SensL's B-Series manufacture and is repeated for all production SiPM sensors that are produced.

Following the PCM test, all SiPMs are tested at CP using tester hardware that has been developed for SiPM sensor testing. For B-Series SiPM sensors, the breakdown voltage, dark current, and optical response to short wavelength light are critical parameters that are measured on the SiPM product die. SensL has established a wafer probe flow that measures every die on the wafer under dark and illuminated conditions, with appropriate optical filters, so that the response of the SiPM sensor under broadband and short wavelength conditions is measured. Die with failing characteristics are identified on electronic wafer maps, and only those die that pass all of the quality and performance screens are subsequently assembled as packaged sensors. Further details of the key parameters that are measured at CP are provided in Sections 7.5.1 and 7.5.2.

7.5.1 BREAKDOWN VOLTAGE

The breakdown voltage is defined as the value of the voltage intercept of a parabolic line fit to the current versus voltage curve, and in the CP wafer test flow, this parameter is calculated from the measurements of the dark current at several bias points. Figure 7.15 shows the distribution of breakdown voltage values as measured from several production lots that are obtained during the qualification of a single die-size product. For the breakdown voltage variation plot shown here, a total sample size of ~204k SiPM sensors of the 30035 type was included. A mean breakdown voltage of 24.69 V was measured. The measured values are seen to be tightly distributed with a standard deviation of 73 mV, and all die have breakdown voltage values within ±0.25 V of the mean.

The uniformity of the breakdown voltage is believed to be due to the tight process control and design of the SensL SiPM. Similar data were obtained for 60035 SiPM at the wafer level.

7.5.2 DARK CURRENT

The dark current is a key parameter for the SiPM and is measured in the CP flow at several bias points below and above the breakdown voltage. Die with dark currents

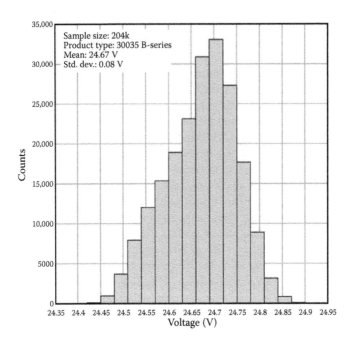

FIGURE 7.15   Histogram of the breakdown voltage distribution for 30035 SiPM. Data are from ~204k production sensors.

that are outside specification are screened out and inked electronically so that they are not assembled into a package. Figure 7.16 shows the tight distribution of dark current values for the 30035 SiPM. For Figure 7.16, all SiPMs were measured at an overvoltage that is equal to 2.5 V.

Figure 7.16 shows the tight distribution of the SiPM dark current and is within product specification. Additional work is currently underway to reduce the dark count for future SiPM sensors through process improvements.

7.5.3   OPTICAL CURRENT

To ensure the quality of the SiPM, the output current from the sensor die is measured in the CP flow under tightly controlled illumination conditions at a number of overvoltages. For the measurements shown here, all SiPMs were measured at a fixed constant voltage of 29.5 V with a uniform and calibrated blue light source. The output current measurement allows an integrated assessment of the sensor gain and PDE. Any variation in gain or PDE has the potential to impact the optical current. By measuring all die on the wafer, it is possible to develop a full understanding of the optical uniformity of the SiPM. Figure 7.17 shows the distribution of output current values as measured from several production lots, with a total sample size of approximately ~142k SiPM sensors of the 30035 type. This measurement procedure was only established during the latter portion of this work, and therefore the size of the data set is smaller than the data set that represented the breakdown voltage and dark current measurements.

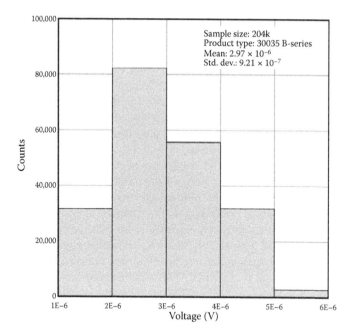

FIGURE 7.16 Histogram of dark current data from 30035 SiPM. Sample size is ~204k, and all SiPMs were measured at an overvoltage equal to 2.5 V.

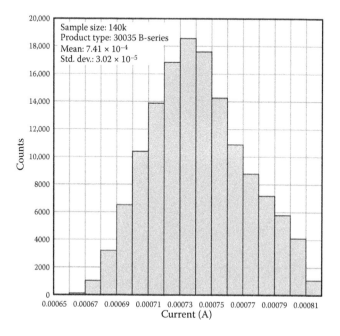

FIGURE 7.17 Optical uniformity at a constant voltage equal to 29.5 V.

In a high-volume application, it is desirable to be able to operate all of the SiPMs in a system at the same voltage. Using a fixed voltage in this test, as opposed to a fixed overvoltage, fully demonstrates the uniformity of the SiPM. By measuring with a fixed voltage, the effect of breakdown voltage variations from die to die, wafer to wafer, and lot to lot were included, and variations in parameters such as film thickness, doping level, and gain are integrated into the final result that is shown. The measured distribution illustrates the tightly controlled uniformity characteristics of the B-Series SiPM manufacturing flow, with uniformity in the range of +/−10% for optical currents at these bias conditions.

While the uniformity of the SiPM sensors has been shown to be extremely good, it is useful to investigate the major cause of the uniformity variation in production SiPM. This can be accomplished by plotting the breakdown voltage of each SiPM from a wafer lot to the optical uniformity that was measured at 29.5 V. This is shown in Figure 7.18 for over 50k die from a single wafer lot. The die type was 30035, and all data are plotted in the figure. This shows a correlation coefficient of 0.87 to a linear fit of the data indicating a strong link between the breakdown voltage and the optical uniformity that is measured. This demonstrates the advantages of tight process control on the final uniformity that can be achieved using the SiPM that is manufactured in this manner.

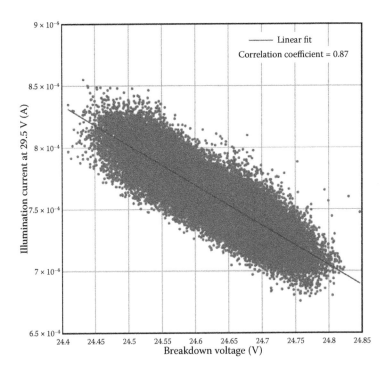

FIGURE 7.18 Correlation plot generated from wafer-level data showing the illumination current at 29.5 V against the measured breakdown voltage from over 50k 30035-type SiPM.

7.6   PACKAGING

The packaging of the sensor plays a large role in the overall performance of the system and one that is not often considered in assessing sensor performance. Multiple factors such as wavelength, allowable operating temperature range, soldering conditions, and reliability vary depending on the package type and construction. In this section, we will describe the main types of packaging that are used for SiPMs that are suitable for radiation detectors. A key criterion for many radiation detectors is the need for a large-area optical sensor to read out a large scintillator or a large scintillator array. This necessitates packages that have the ability to be closely tiled together to form large-area sensors with single readouts or sensor arrays with multiple SiPM readouts. The package types discussed in this section are all suitable for use in large-area arrays and can be pixelated or summed as required by the user.

There are three major packaging types that are suitable for use with scintillation crystals in a radiation detector. These package types are the following:

1. Poured epoxy-coated die in a ceramic or laminate carrier
2. Clear micro leadframe package (MLP)
3. Wafer-scale-packaging through-silicon via (TSV)

Each of these package types has positive and negative aspects, which should be understood during a product design process. To assist in understanding the variations between packages, the following criteria are critical to understand and can be used to evaluate package suitability for an application. The criteria used to decide upon a package type are the following:

- *Array fill factor*: The ratio of optically active area to peripheral area. The fill factor is important when creating arrays of SiPM, which are typically available in 1-, 3-, and 6-mm^2 sizes. This parameter is not critical if a single sensor can be used alone in a detector. For applications requiring large-area coverage, a high-package fill factor is critical in order to capture as many incident photons coming from the scintillation crystal.
- *Optical transmission*: The transmission characteristics of a package vary depending on the package type that is chosen. Compounding this problem is that the optical transmission can be wavelength dependent. As shown in Section 7.8.1, the peak emission wavelength varies between crystal types. Therefore, it is important to make sure that the package does not limit the light collection from the scintillation material.
- *Operating conditions*: The temperature that a package can physically operate in depends on the package construction and material.
- *Reliability*: The various package types used for sensors have different failure mechanisms, which lead to different inherent reliability.
- *Service life*: The service life or wear-out time of the sensor will be determined primarily by the degradation in transmissivity of the package material.

- *Uniformity and reproducibility*: The uniformity of the package surface is important to maintain optimum crystal coupling to an SiPM. The reproducibility of the package is important so that consistent results can be obtained in large-area array applications.
- *Cost*: The cost of a package must be understood and is important for high-volume applications. Low cost is obviously desirable if the performance and the operating life criteria can be met.

Sections 7.6.1 through 7.6.3 will review each of the three main package types with respect to the criteria that have been outlined here.

7.6.1 POURED EPOXY PACKAGES

This category covers a broad range of packaged parts that all use poured epoxy as the covering layer over the SiPM. The epoxy layer provides protection to the SiPM surface and prevents damage to wire bonds. For large-area arrays, the fabrication process entails attaching multiple die to a ceramic or PCB carrier using a die attach epoxy. Following die attach and cure, the contact pads on the SiPM are wire that is bonded to the substrate. For the creation of scalable large-area arrays with uniform performance and wiring trackout, we will only discuss packages here in which the wire bonds are connected directly to the substrate beside the SiPM sensor. Stitch bonding between die can be used to create higher-density arrays; however, this leads to variable SiPM trackout lengths on large arrays and is not scalable to a large area. Stitch bonding is therefore not discussed here in detail. Following wire bonding, the SiPM and wire bonds are protected by a poured coat of clear epoxy, which is subsequently cured. An example of the completed package is shown in Figure 7.19.

This package has external dimensions of 57.4 × 57.4 mm and contains 64 6 × 6-mm^2 SiPMs. The SiPMs are placed with a dead space 0.80 mm between the die, which allows for wire bond access. The fill factor that can be achieved with this type of package is shown in Figure 7.23.

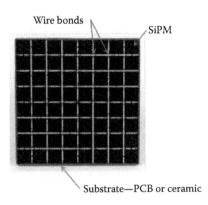

FIGURE 7.19 Epoxy-coated ceramic carrier SiPM array. This example is an 8 × 8 array of 6 × 6-mm^2 SiPM that is mounted on a PCB substrate.

Poured epoxy packaging has some advantages. The robust substrate enables manual handling during processing. This and the low-capital investment required make it suitable for high-mix, low-volume manufacturing. The fill factor can also be good because the common substrate enables a minimum distance between dies (generally 0.020 mm) for wire bonding. Additionally, this package type can be made with no ferromagnetic material, which makes it a candidate for use in magnetic resonance (MR) imaging applications where ferromagnetic materials will degrade MR images.

The most significant disadvantages of this product are related to the epoxy itself. Generally, this is done by mixing two parts epoxy and pouring it onto the die surface. Poured epoxies will cure at room temperature, though elevated temperature is often used to decrease the hardening time. It is difficult to control this process, and inconsistent epoxy material properties result, which affect product performance in the following ways:

- *Inconsistent PDE due to variations in optical transmission, particularly at short wavelengths.*
- *Degradation in optical transmission during operating life ("yellowing" of the organic epoxy).*
- *Nonuniformity of adhesion to the die surface.* This will result in delamination during the operating life and the consequent loss of transmission or failure of the wire bond interconnects.
- *Nonuniformity and high values of mismatch between thermal expansion coefficient of the epoxy and that of the substrate and wire bonds, leading to failure of wire bonds.*
- *Poured epoxy products require a package sidewall to contain the epoxy.* Consequently, there is always dead space at the edges of the array. This adds dead space between arrays that are placed beside other arrays and limits the total fill factor.
- *Connection to the array must also be made to the back of the substrate.* This increases the PCB traces and connections between the sensor and the subsequent readout electronics, which can degrade the performance when compared to the other package technologies that are discussed here.

A summary of the most important points related to the poured epoxy package is detailed in Table 7.2.

7.6.2 CLEAR MLP

The clear MLP process entails attaching the SiPM to a large leadframe with die attach epoxy, wire-bonding the SiPM contact pads to the leadframe terminals, molding in clear compound, and sawing into individual packages. Since the SiPMs are mounted on to a much larger leadframe, a fully automated process can be used in their manufacture. A drawing of a clear MLP SiPM is shown in Figure 7.20a. This figure details the SiPM placement on the leadframe, wire bond placement, and clear molding compound, which encases the SiPM. Examples of three different-sized, clear MLP–packaged SiPMs are shown in Figure 7.20b. This figure shows a

TABLE 7.2

Summary of Comparison Points for SiPM Array Package Options

Requirement	Poured Epoxy	Clear MLP	TSV
Array fill factor	Near optimum	Limited by peripheral leadframe	Optimum
Optical transmission	Poor	Good	Best
Operating conditions	0°C–40°C	–40°C–85°C	–40°C–85°C
Reliability	Manual processing: reduced reliability	Good	Best
Service life	Yellowing of potted epoxy is not well controlled	Good	Best
Uniformity and reproducibility	Poor	Good	Good
Cost	Low cost for laboratory use—not recommended for use in volume arrays	Lowest high-volume sensor cost—low array assembly costs	Highest high-volume sensor cost—low array assembly costs

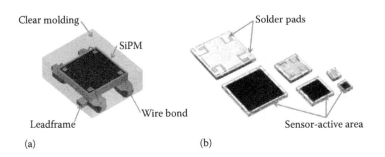

(a)   (b)

FIGURE 7.20   (a) Clear MLP drawing. (b) Images of the top and back side of SiPM 1 × 1-, 3 × 3-, and 6 × 6-mm² SiPM-packaged sensors.

6 × 6-, 3 × 3-, and 1 × 1-mm² SiPM sensor. The back contact of the sensor is shown at the top of Figure 7.20b.

The epoxy in a clear MLP is molded and cured at temperatures between 140°C and 160°C. The material and mold process are more controllable than the mixed and poured epoxy that was described in Section 7.6.1 on epoxy-covered substrate packages. The MLP SiPMs have the following key benefits over poured epoxy SiPMs:

- Better control of optical properties and thermal expansion or contraction of the material during operating life.
- A flatter, more uniform surface and optical interface allow better coupling to scintillation crystals.

- Clear MLP SiPM sensors can be surface-mounted directly to the readout PCB, and it is possible to minimize the PCB trace length and reduce the number of connectors in a large-area array.
- A moisture sensitivity level (MSL) of 3 can be obtained, which allows 168 h of ambient humidity exposure before the part needs to be soldered to a PCB.
- Clear MLP sensors can be soldered using standard reflow solder conditions at 260°C.
- MLP materials and processing are suited to high-volume manufacturing and provide the lowest-cost product compared to any other package that is suitable for arrays.
- Arrays can be fabricated with minimal dead space at the edge. This allows arrays to be tiled at a constant sensor pitch.

Examples of four different types of array, manufactured using clear MLP sensors, are shown in Figure 7.21. In all cases, these arrays were assembled on a PCB with 0.2 mm of dead space between the sensors. The largest-size array is the 8 × 8 array of 6 × 6 mm², which is 57.4 mm per edge. For this array, the sensors were placed on a 7.2-mm pitch. The pitch comprises a 6-mm sensor, 1 mm of package dead space, and 0.2 mm of dead space between packages. The 1 × 1-mm² sensor is packaged in a 1.5 × 1.8-mm² package enabling a high fill factor.

In addition to the performance and reliability benefits of the clear MLP, the flexibility to define the array size through PCB layout and the ability to use high-volume PCB manufacturing mean that the clear MLP is considered a superior package type to those utilizing poured epoxy.

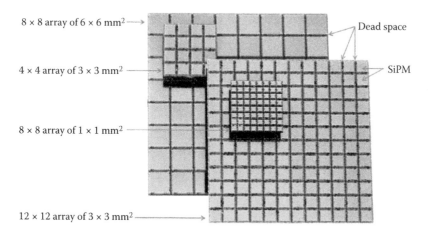

FIGURE 7.21 Example of four different large-area arrays of clear MLP array sensors mounted on PCB. The dead space between clear MLP sensors is 200 µm.

7.6.3 TSV

Both poured epoxy and clear MLP require packaging of die after they have been processed. TSV packaging, however, is a wafer-level package and is made while the die are still on the wafer and have not yet been singulated. A true wafer-scale TSV process entails bonding a glass substrate to the top of the silicon wafer and back-grinding the silicon to a thin layer. Vias are etched in the silicon, and metal is deposited to make contact with the terminals in the die. Solder bumps are attached to the deposited metal on the back side of the sensor. The wafer is then sawn into individual die and can be placed onto a PCB substrate with a similar process to that which is used to manufacture clear MLP arrays. An example of a TSV-packaged SiPM is shown in Figure 7.22a. The corresponding cross section of an SiPM TSV sensor is shown in Figure 7.22b.

The performance of a TSV-packaged SiPM brings the PDE closer to the bare-die PDE curve, as shown in Figure 7.23. This is due to the fact that the TSV glass is more transmissive at short wavelengths than the organic materials that are used in poured epoxy and clear MLP products. The additional benefits of a TSV-packaged SiPM are as follows:

- *Transmission in glass is stable during operating life.* Yellowing seen in organic compounds does not occur in glass, and TSV products have the longest operating lifetime.
- *There are no wire bonds, so each TSV component will be the same size as the sensor.* This optimizes the fill factor allowing the highest-density arrays to be produced. This high fill factor is shown in Figure 7.24.
- *The failure mechanisms associated with wire bonds and epoxy are absent.* Therefore, TSV products will have the highest reliability of any SiPM package.
- *TSV packages are MSL 3 compatible, reducing costs at PCB contract manufacturers.*
- *TSV processing is inherently conducive to tight control and high volume.*

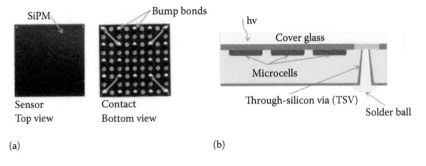

FIGURE 7.22 (a) TSV SiPM example showing top sensor side and bottom contact side. (b) TSV cross-section example.

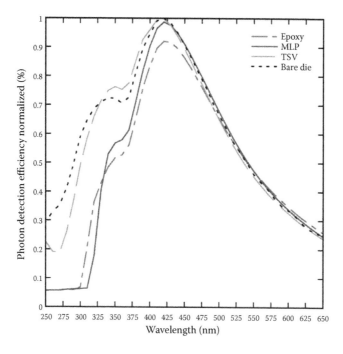

FIGURE 7.23 Comparison of PDE versus wavelength measured on bare die, poured epoxy, clear MLP, and TSV package types.

- *TSV can be placed into arrays with low dead space between sensors.* Due to excellent wafer saw tolerances, the dead space between TSV die in arrays can be 0.2 mm. This improves the overall TSV package fill factor compared to clear MLP.

The key disadvantage of a TSV package is a slightly increased cost as compared to clear MLP. It is up to the user to decide if the advantages outweigh a sensor cost increase in their product design.

7.6.4 SUMMARY INFORMATION

The following figures and tables summarize the various packages that were discussed in this section. Figure 7.23 compares the PDE that is measured versus wavelength for four differently packaged 3×3-mm^2 SiPM sensors. All data have been normalized to allow easy comparison between package types. To provide a baseline of the core performance, the bare die curve was obtained with an SiPM that is mounted directly to a substrate with no poured epoxy over the surface area. It therefore represents the PDE of the sensor. The epoxy curve represents the typical PDE that is measured with poured epoxy products; this is characterized by a reduced peak PDE due to transmission losses in the epoxy. The clear MLP package has better peak PDE than the poured epoxy but does have a transmission floor at 320 nm due to the clear molding compound. The TSV package curve demonstrates the best performance at the peak wavelength and

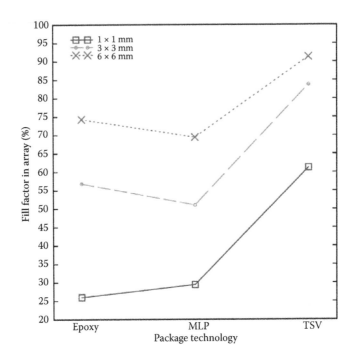

FIGURE 7.24 Comparison of fill factor between poured epoxy, clear MLP, and TSV package types when placed in an array.

greatly improved performance into the ultraviolet (UV) range. The TSV package curve demonstrates the best performance at the peak wavelength and greatly improved performance into the UV. The TSV package most closely approximates the bare die curve and represents the highest-performance package for the highest PDE.

A comparison of the fill factors of various package types was performed using the detailed knowledge of each SiPM sensor size: 1×1, 3×3, and 6×6 mm^2 combined with typical design rules for contract manufacture placement of sensors into arrays. In all cases, the TSV package provides the highest package fill factor that is possible when creating large-area arrays due to the low dead space.

The comparison of the three package technologies is summarized in Table 7.2.

7.7 SiPM RELIABILITY

In this section, the reliability stress assessment completed on SiPM will be discussed. Additionally, the justification for the adoption of industry standard reliability stress assessment flows from Joint Electron Devices Engineering Council (JEDEC) will be reviewed.

7.7.1 ISSUES OF SIPM RELIABILITY ASSESSMENT

SiPM sensor reliability is a key requirement for high-volume applications. At present, there is a lack of industry-accepted standards for the reliability assessment of

SiPM sensors. This is believed to be due to the varied and complex manufacturing process that is required for optical sensors. This is contrasted to the integrated circuit industry where there are strong standards that are widely adopted and used for the assessment of the reliability of products. As the sensors discussed in this work are developed using CMOS in high-volume fabrication facilities, it is applicable and desirable to follow standards that are adopted by the integrated circuit industry. A major publisher of standards used by the integrated circuit industry is JEDEC,[25] which freely publishes test standards that can be adopted by manufacturers to assess product reliability. For the reliability assessment of SiPMs, the JEDEC standards have been adopted.

7.7.2 RELIABILITY ASSESSMENT OF SiPM

All stress and test steps were carried out as per JEDEC standard conditions. The JEDEC standard for the main reliability test procedure can be obtained from JEDEC.[25] To assess the silicon reliability, SensL used a combination of standard production packages that are available, including SensL clear MLP and metal can packages with sealed lids containing an optical window. The use of metal can packages was a requirement for the high-temperature stress tests that are designed to test the reliability of the silicon, independent of the package material. For the assessments, the packaged die were obtained typically from three separate silicon fabrication lots and packaged as required. All tests were made in accordance to the relevant JEDEC standards that are referenced in Table 7.3. Of significant note from Table 7.3 were the high-temperature operating life tests that were completed to 1000 h with an additional higher endurance stress to 2000 h. This represents the longest high-temperature stress to date on a statistically relevant sample set of SiPM, which the authors have seen reported.

All SensL SiPM sensors passed the tests. An example of the PDE measured before and after a high temperature, an operating-life stress test, is shown in Figure 7.25. No change in performance was observed.

As can be seen in the table, further stress data were taken to assess product reliability, including temperature cycling, moisture and humidity stressing, and high-temperature storage. The excellent results demonstrate that the SiPM manufacturing process is robust and enables the production of reliable, high-quality sensor components.

For high-volume applications, the reliability of the product is a key concern. It is believed that the adoption of reliability assessment standards used by integrated circuit manufacturers provides the most comprehensive method to assess SiPM product reliability. The standard methods outlined by JEDEC have been applied to B-Series SiPM from SensL, with all parts passing standard reliability assessment flows. The adoption of integrated circuit industry standard reliability test flows is considered of vital importance for the adoption of the SiPM in high-volume applications.

7.8 SCINTILLATION CRYSTAL READOUT

The focus of this section is on the use of scintillation crystals for the detection of gamma rays. Such crystals are used for a variety of applications that are related to

TABLE 7.3
Reliability Stress Program for SiPM

Test	Objective	Required Condition	Lot Size	Duration/Acceptance	Status
High-temperature operating life	Junction stability	Ambient temperature = 125°C; bias = 30 V	3 lots of 77 units	1000 h/no change in any parameter > 10%	100% passed
High-temperature operating life	Junction stability over longer stress time	Ambient temperature = 125°C; bias = 30 V	256 units	2000 h/no change in any parameter > 10%	100% passed
High-temperature operating life	Package stress to examine chemical stability (e.g., discoloration of package)	Ambient temperature = 85°C; bias = 27 V	1 lot of 77 units	1000 h/no change in any parameter > 10%	100% passed
Unbiased, highly accelerated stress	Package stress to examine delamination, transmission loss, and wire bond failure	110°C, 85% relative humidity; Passive, no bias	3 lots of 25 units	264 h/no change in any parameter > 10%; no critical package delamination	100% passed
Temperature cycling	Package stress to examine delamination, transmission loss, and wire bond failure	−40°C–85°C cycle, 15-sec transition, 15-min dwell time; Passive, no bias	3 lots of 77 units	500 cycles/no change in any parameter > 10%; no critical package delamination	100% passed
High-temperature storage test	Package stress to examine chemical stability (e.g., discoloration of package)	504 h at 125°C; Passive, no bias	3 lots of 25 units	504 h/no change in any parameter > 10%	100% passed

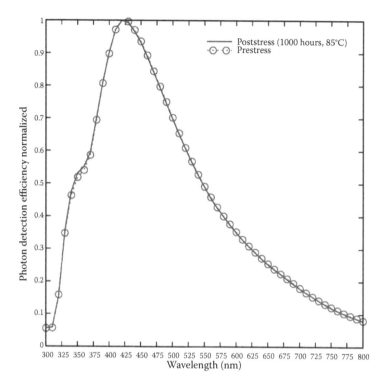

FIGURE 7.25 Measurement of PDE pre- and poststress. Stress condition is 85°C for 1000 h.

medical imaging and radiation detection. Both markets use the combination of a scintillator and an optical sensor to detect gamma rays.

7.8.1 GAMMA-RAY DETECTORS

A gamma-ray detector must detect, amplify, and output an electrical signal that accurately conveys the gamma-ray energy and, in some cases, the photon arrival time. The most common type of gamma-ray detector relies on a combination of a scintillation crystal and an optical sensor, as shown in Figure 7.1. A variety of crystals are available that can be used to convert the gamma radiation into optical photons, and the work of Dorenbos[16] shows a useful timeline of their development and performance limits for various crystal types. Some of the most popular scintillators, such as LaBr3, LYSO, bismuth germanium oxide (BGO), and CsI(Tl), which are relevant to PET, and radiation detection will be considered in this work. These crystals have a variety of primary characteristics such as light yield, density, emission wavelength, and decay time, which directly impact the overall performance of the crystal and optical sensor assembly. Typically, to optimize the performance of a gamma-ray detector, the goal is to maximize the number of optical photons that are detected from the scintillator while minimizing the amount of noise coming from the optical sensor.

Figure 7.26 shows the normalized light yield for a variety of standard scintillation crystals that are used for radiation detection. The data for Figure 7.26 were obtained directly from the datasheets that are available for each crystal type from Saint-Gobain[17] and in agreement with published data that were found in the extensive scintillation crystal database that was made available online by Derenzo.[18] The photon emission has been digitized from manufacturer datasheets and normalized according to the peak light yield, which is expressed in photons per kiloelectronvolt. A high light yield and an emission spectrum matched to the sensitivity of the SiPM will typically improve detection performance.

Another parameter that plays an important role in the performance of radiation detectors is the emission decay time. The typical decay times for the crystals discussed in this work are shown in Table 7.4. The decay time of the crystal determines the amount of time that is required for measurement and the amount of noise that will be integrated during the signal acquisition. Longer measurement times lead to increasing amounts of noise that is included in the recorded signal.

The refractive index of the crystal is also important as this will determine the light coupling between the crystal and the sensor. Mismatched refractive indices lead to Fresnel losses of photons transitioning from the scintillation crystal to the optical sensor. Index-matching coupling material should always be used between an SiPM sensor and the scintillation crystal for optimum light collection.

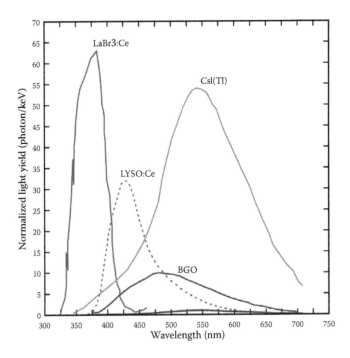

FIGURE 7.26 Crystal emission light yield from common scintillation crystals used in radiation detection.

TABLE 7.4

Typical Crystal Information from Saint-Gobain[17] and Energy Resolution Obtained in the Literature from Dorenbos[32]

Scintillation Crystal	Emission Peak λ (nm)	Light Yield (Photons/keV)	1/e Decay Time (ns)	Refractive Index at λ (Peak)	Energy Resolution (%) at 662 keV
LaBr3:Ce	380	63	16	1.9	~2.6
LYSO:Ce	420	32	41	1.81	~8
BGO	480	10	300	2.15	~10
CsI(Tl)	550	54	1000	1.79	~5.5

The scintillation crystal emission and the decay time of the scintillator combined with sensor sensitivity and noise play a large role in radiation-detector performance optimization. Additional scintillation crystal parameters such as density, thermal expansion coefficients, hygroscopic nature, and cost will also play a role in determining the suitability of a detector for a given application, but these are outside of the scope of this work on SiPM sensors.

7.8.2 ENERGY RESOLUTION AND COINCIDENCE TIMING MEASUREMENTS

Applications involving scintillation materials fall into two main application areas: (1) those that require fast timing and (2) those for spectroscopy requiring optimum energy resolution. These two application areas have slightly diverging requirements in the scintillating crystals that are used and in the optimal operating voltage settings for the sensor. For fast-timing applications, it is required to have a high light yield scintillator with a fast rise and short decay time. For applications requiring the best (lowest) energy resolution, the nonproportionality of the crystal is a driving factor to obtain the lowest energy resolution. Energy resolution values, which have been obtained in the literature, are summarized in Table 7.4.

For the best energy resolution measurements, the maximum of the sensor PDE should be matched as closely as possible to the peak wavelength of the scintillator. The energy resolution percentage (full width half maximum [FWHM]) is related to both the sensor performance and the crystal. A good review of this is found in the work by Dorenbos[16] and Grodzicka,[27] which determines the energy resolution through the following equation:

$$R^2 = R_{in}^2 + R_{np}^2 + R_{stat}^2 \tag{7.12}$$

where R_{in} is related to the crystal inhomogeneity from crystal quality, wrapping variation, and crystal finish differences; R_{np} is due to the nonproportional response of the scintillator; and R_{stat} is related to the photon detection statistics and the excess noise of the SiPM. R_{stat} is given by the following equation,[22] which has been modified to include the dark count rate in one formula:

$$R_{\text{stat}} = 2.35 * \frac{\sqrt{\text{ENF} * \left(N_{\text{dp}} + N_{\text{dc}}\right)}}{N_{\text{dp}}} \qquad (7.13)$$

In this equation, N_{dp} is the number of detected photons, and N_{dc} is the number of dark counts during the measurement integration time. ENF is the excess noise factor. An ENF of 1 is a perfect detector, while increasing the ENF degrades R_{stat}. The ENF for the SiPM is dominated by crosstalk and afterpulsing. These unwanted effects increase variability in the output of the SiPM. When used in a detector, increasing the overvoltage will increase the PDE and therefore the number of detected photons, N_{dp}; however, crosstalk and afterpulsing will also increase, which can degrade the energy resolution that is obtained. For typical PET applications, looking at signal levels from 511-keV gamma rays, R_{stat} is the predominant factor, whereas R_{dcr} dominates measurements at lower energies.

For coincidence timing measurements, the scintillating crystal emission should be short, and the PDE and rise time of the sensor should be high. Crosstalk, afterpulsing, and dark count rates are considered secondary factors until high overvoltages are reached. This operating point is best determined experimentally as it depends on the scintillator, sensor, and readout method. Crystals for use in fast coincidence timing measurements emit light in the blue to the UV portion of the spectrum.[16] The SiPM sensors shown in this work have been optimized to allow the detection of these short-wavelength emission crystals.

7.8.3　SiPM Scintillation Crystal Performance

To demonstrate the performance of the SensL SiPM in an application environment, the sensors were coupled with cerium-doped lutetium yttrium orthosilicate (LYSO:Ce) scintillation crystals. This is similar to the system configuration in a time-of-flight positron emission tomography (ToF-PET) system. A ToF-PET system is defined as a PET system in which the coincidence resolving time (CRT) needs to be determined to 500 ps or better. This is a subtle distinction but it is a common one used in this market to separate non-ToF from ToF-PET systems. When the SiPM sensor is combined with the scintillation crystal, it forms a gamma-ray detector. It is desired to demonstrate a minimum energy resolution and minimum CRT of the pair 511-keV gamma rays that are produced in electron–positron annihilation. LYSO is well matched to the sensor PDE that was presented in Figure 7.7 as the output peak at 420 nm matches the photon detection peak of the sensor. The measurement setup for energy resolution and CRT is shown in Figure 7.27. This shows the crystals in *head-on* configuration, which more closely mimics the way crystals are positioned in a standard PET system. The crystals are directly mounted on the SiPM using coupling grease (BC-630),[17] and amplifiers are used to amplify the output voltage level to an appropriate range for the Wavecatcher Universal Serial Bus (USB)

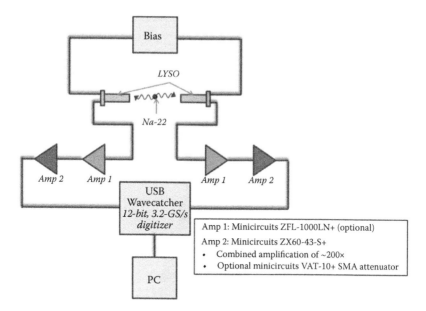

FIGURE 7.27   Example of the energy resolution and CRT experiment setup used in this work. SMA, subminiature version A connector.

digitizer.[28] A variety of amplifier[24] combinations can be chosen to suit most crystal testing needs.

7.8.4   CRT

CRT is an important ToF-PET system measurement parameter. A low CRT allows a more accurate determination of emission source position through triangulation. For this work, the CRT was evaluated by determining the arrival time of coincident 511-keV photon pairs at a corresponding pair of scintillation-based detectors. For this measurement, a ^{22}Na 511-keV source was placed between two facing $3 \times 3 \times 20$-mm^3 LYSOs that are coupled to SiPM sensors, positioned head on, and the resulting electrical signal from the detectors was amplified and recorded with a high-speed digitizer (USB Wavecatcher 12 bits, 3.2 GS/s). In analyzing the resulting pairs of signal traces, energy filtering was performed at 9% of the peak value-selecting signals that the charge integral matched with the peak in the charge spectrum that is attributed to 511 keV; time walk correction was performed removing the correlation between the energy difference and time difference of scintillator pairs. In addition, the optimal choice of leading-edge threshold to time-stamp each of the electrical signals further optimized the CRT value. Figure 7.28 shows typical CRT histograms that are measured on the fast output of 30035 SiPM. A value of 225 ± 2 ps was obtained. Similar results have been shown in the literature.[19,21]

Measurements on standard anode–cathode signals were also performed, and a value of 306 ± 3 ps was obtained. The fast output provides superior results due to its fast signal rise time, low output capacitance, and low parasitic impedance.

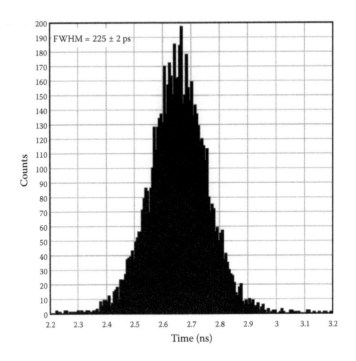

FIGURE 7.28   CRT measured for sample pairs of 30035 SiPM.

7.8.5 ENERGY RESOLUTION

In energy resolution measurements, the detector principle is that of a proportional counter in that the number of scintillation light photons are proportional to the energy of the gamma photon penetrating the crystal. Integrating the electric signal from the sensor during each scintillation event and then binning the result yields a spectrum of counts that is proportional to the gamma photon energy.

To determine the energy resolution for a typical PET setup, a $3 \times 3 \times 20\text{-mm}^3$ LYSO crystal was coupled to a 30035 SiPM sensor using BC-630 silicone optical grease from Saint-Gobain. The SiPM was biased at 5 V over the breakdown voltage, and the signal from the SiPM was amplified, and traces were recorded using a high-speed digitizer.

The energy resolution spectrum for 30035 SiPM is shown in Figure 7.29. This measurement was made from the fast-output signals and was similar to measurements that are taken on the standard anode–cathode output. When coupled to a relatively bright and fast crystal such as LYSO, the signal from the fast output is sufficient to accurately represent the signal pulse, and an accurate energy resolution plot can be obtained. For this measurement, the energy resolution was limited by the crystal proportionality, statistical uncertainty in photon counting by the sensor, and electrical noise in the sensor and associated electronics.

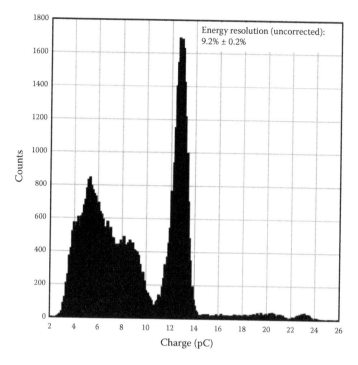

Energy resolution (uncorrected):
9.2% ± 0.2%

FIGURE 7.29   Charge spectrum FWHM measured on fast output for 30035 SiPM.

Additional work on measurements using CsI(Tl) can be found in the work of Becker[29] who details a compact and portable gamma-ray spectrometer measurement system, which achieves 5.66% energy resolution.

7.9   SUMMARY

This work has demonstrated the operation of SiPM that is developed for high-volume CMOS fabrication. The review started from the basic physics of the operation of SiPM and linked various key operating parameters, such as the breakdown voltage and wavelength sensitivity, to the physical structure of the sensor. The main SiPM parameters were reviewed including PDE, dark count rate, gain, crosstalk, afterpulsing, and dynamic range. The key highlights show that a PDE of 41% and low dark count rates of <100 kHz can be obtained with low afterpulsing and crosstalk rates. The manufacturability and high-volume production of SiPM were demonstrated through wafer-level electrical and optical testing of over 204k SiPM sensors demonstrating excellent V_{br}, dark current, and optical uniformity. A review of the packaging requirements and suitable packages for large-area arrays was presented demonstrating the key parameter trade-offs such as PDE, fill factor, and cost. It was shown that clear MLP- and TSV-packaged sensors have evident advantages for high-volume array applications. For the reliability assessment of the SiPM industry, standard integrated circuit-testing procedures were adopted. The tests are believed to provide a comprehensive testing regime for SiPM and should be used for SiPM product qualification. The use of SiPM for scintillation crystal readout was reviewed theoretically, and a clear demonstration of the high performance that can be obtained was shown with LYSO:Ce CRT of 225 ps and energy resolutions of 9.2% that are obtained experimentally.

REFERENCES

1. R. H. Haitz. 1964. Model for the electrical behavior of microplasma. *Journal of Applied Physics*, vol. 35, no. 5, pp. 1370–1376, May.
2. R. H. Haitz. 1964. Microplasma interaction in silicon p-n junctions. *Solid-State Electronics*, vol. 7, pp. 439–444.
3. R. H. Haitz. 1965. Mechanisms contributing to the noise pulse rate of avalanche diodes. *Journal of Applied Physics*, vol. 36, no. 10, pp. 3123–3131, October.
4. R. H. Haitz. 1965. Studies on optical coupling between silicon p-n junctions. *Solid-State Electronics*, vol. 8, pp. 417–425.
5. W. G. Oldham et al. 1972. Triggering phenomena in avalanche diodes. *IEEE Transactions on Electron Devices*, vol. 19, no. 9, pp. 1056–1060, September.
6. R. J. McIntyre. 1966. Multiplication noise in uniform avalanche diodes. *IEEE Transactions on Electron Devices*, vol. ED-13, pp. 164–168.
7. R. J. McIntyre. 1973. On the avalanche initiation probability of avalanche diodes above the breakdown voltage. *IEEE Transactions on Electron Devices*, vol. 20, no. 7, pp. 637–641, July.
8. S. Cova et al. 1981. Towards picosecond resolution with single-photon avalanche diodes. *Review of Scientific Instruments*, vol. 52, no. 3, pp. 408–412, March.
9. R. G. W. Brown et al. 1986. Characterization of silicon avalanche photodiodes for photon correlation measurements. 1: Passive quenching. *Applied Optics*, vol. 25, no. 22, pp. 4122–4126, November.
10. R. G. W. Brown et al. 1987. Characterization of silicon avalanche photodiodes for photon correlation measurements. 2: Active quenching. *Applied Optics*, vol. 26, no. 12, pp. 2383–2389, June.
11. A. B. Kravchenko et al. 1978. A linear array of avalanche MIS photodetectors. *Soviet Journal of Quantum Electronics*, vol. 8, no. 11, pp. 1399–1400, November.
12. A. B. Kravchenko et al. 1978. Feasibility of construction of a pulsed avalanche photodetector based on an MIS structure with stable internal amplification. *Soviet Journal of Quantum Electronics*, vol. 9, no. 9, pp. 1086–1089, September.
13. V. Saveliev et al. 2000. Silicon avalanche photodiodes on the base of metal-resistor-semiconductor (MRS) structures. *Nuclear Instruments and Methods in Physics Research A*, vol. 442, pp. 223–229.
14. D. Bisello et al. 1995. Metal-resistive layer-silicon (MRS) avalanche detectors with negative feedback. *Nuclear Instruments and Methods in Physics Research A*, vol. 360, pp. 83–86.
15. V. Golovin et al. 2000. Limited Geiger-mode microcell silicon photodiode: New results. *Nuclear Instruments and Methods in Physics Research A*, vol. 442, pp. 187–192.
16. P. Dorenbos. 2010. Fundamental limitations in the performance of Ce3+, Pr3+, and Eu2+ activated scintillators. *IEEE Transactions on Nuclear Science*, vol. 57, pp. 1162–1167.
17. Saint-Gobain. Available at http://www.crystals.saint-gobain.com/, accessed June 2014.
18. S. Derenzo. Scintillation properties. Lawrence Berkeley National Laboratory. Available at http://scintillator.lbl.gov/, accessed June 2014.
19. Silicon refractive index. Available at http://refractiveindex.info, accessed June 2014.
20. J. Y. Yeom et al. 2013. Fast timing silicon photomultipliers for scintillation detectors. *IEEE Photonics Technology Letters*, vol. 25, pp. 1309–1312.

21. S. Dolinsky et al. 2013. Timing resolution performance comparison for fast and standard outputs of SensL SiPM. *IEEE Nuclear Science Symposium and Medical Imaging Conference (NSS/MIC)*, Seoul, Korea, October 27, 2013.

22. Moszynski et al. 2009. A comparative study of silicon drift detectors with photomultipliers, avalanche photodiodes and PIN photodiodes in gamma spectrometry with LaBr3 crystals. *IEEE Transactions on Nuclear Science*, vol. 56, no. 3, pp. 1006–1011.

23. SensL Technologies Ltd. B-Series Datasheet. Available at http://www.sensl.com/downloads/ds/DS-MicroBseries.pdf, accessed June 2014.

24. Mini Circuits Inc. website. Available at http://www.minicircuits.com, accessed June 2014.

25. Joint Electron Devices Engineering Council (JEDEC) Standards Organization. Available at http://www.jedec.org, accessed June 2014.

26. J. T. M. de Haas et al. 2008. Advances in yield calibration of scintillators. *IEEE Transactions on Nuclear Science*, vol. 55, pp. 1086–1092.

27. M. Grodzicka et al. 2013. MPPC array in the readout of CsI:Tl, LSO:Ce:Ca, LaBr$_3$:Ce, and BGO scintillators. *IEEE Transactions on Nuclear Science*, vol. 59, no. 6, pp. 3294–3303, December.

28. Wavecatcher USB Digitizer. Available at http://www.heptech.org/Phocea/file.php?file=Ast/140/fiche_LAL_Wave_Catcher.pdf, accessed June 2014.

29. E. Becker. 2013. *The MiniSpec: A Low-Cost, Compact, FPGA-Based Gamma Spectrometer for Mobile Applications*. MS Thesis, Corvallis, OR: Oregon State University. Available at http://ir.library.oregonstate.edu/xmlui/handle/1957/42365, accessed June 2014.

8 Designing Photon-Counting, Wide-Spectrum Optical Radiation Detectors in CMOS-Compatible Technologies

Edoardo Charbon and Chockalingam Veerappan

CONTENTS

8.1 Introduction 185
8.2 CMOS APDs for Time-Resolved Imaging 186
8.3 Noise Performance in DSM CMOS SPADs 187
8.4 CMOS SPADs in Full Isolation 188
8.5 Trading Off Noise with Sensitivity 190
8.6 Dynamic Performance 192
8.7 Timing Resolution 193
8.8 Trends and Comparisons 193
8.9 Conclusions 196
References 196

8.1 INTRODUCTION

Photon counting in one or in arrays of pixels is useful in a number of applications, from more traditional ones, such as low-light-level and fluorescence imaging, super-resolution microscopy, 3D time-of-flight sensing, and NIR optical tomography [1–6], to niche applications, such as voltage-sensitive-dye (VSD)-based imaging [7,8], particle image velocimetry [9], instantaneous gas imaging [10,11], fluorescence-based time-resolved imaging (both single- and multiphoton, and lifetime, and correlation based) [12–14], and applications based on time domain interferometry, like the Hanbury Brown–Twiss interferometry, for the analysis of microlight and macrolight sources as well as for stellar bodies [15–17]. In this context, and in the context of consumer applications, compactness, weight, and power consumption have prompted a boost to the development of solid-state photon-counting sensors, where photoelectrons are multiplied in single-sided abrupt junctions operating at high electric fields that can sustain impact ionization. These structures are generally based on a technology that is known as Geiger-mode avalanche photodiode or single-photon avalanche diode (SPAD), whereas the first devices of this kind were vertically integrated in dedicated processes, e.g., reach-through avalanche photodiode (RAPD) based on p+-π-p-n [18].

Researchers have gradually switched their attention to planar structures, investigated since the 1970s by Cova and coworkers [19]. Many authors have developed avalanche photodiodes (APDs) both in proportional and Geiger mode using *dedicated* processes, achieving superior performance in terms of sensitivity and noise. A good example is the work of Kindt [20]. The main disadvantage of using dedicated processes is generally the lack of libraries that can support complex functionalities and deep-submicron (DSM) feature sizes, thus limiting array sizes.

Proportional APDs and especially SPADs have recently evolved toward more and more compact devices following Moore's law. As a result, functional devices have emerged in 0.8-μm [3,21,22], 0.35-μm [23–27], 0.18-μm [28,29], and 130-nm [30–32] CMOS processes. With the availability of SPADs in DSM CMOS, it has become possible to implement more and more complex functionalities in short proximity to the detector and smaller pixels; these features are essential in the design of large arrays. Nevertheless, the transition from single-photon detection to imaging, especially over large areas, has required readout schemes that could sustain reasonably high fill factors. Thus, many solutions have been implemented over the years, most notably based on column-parallel readout [33–35] and in-pixel digital and analog counting [36–39].

While in [40], we focused on CMOS-compatible photon-counting devices that are designed for operation in hostile environments that are flooded with radiation of various nature and energy content, in this chapter, we are interested in CMOS-compatible photon-counting pixels that are amenable to integration in massively parallel arrays [41], wherein the main focus is wide-spectrum sensitivity, spanning from NUV to NIR ranges. Sensitivity, in the context of photon-counting devices, is characterized in terms of photon detection probability (PDP), which is the probability that a photon impinging the active area of the sensor is detected and multiplied to generate a count. If fill factor, i.e., the area ratio between active and pixel area, is taken into account, then one refers to PDE. We will also look at the implications of wide-spectrum sensitivity in terms of time resolution, PDE uniformity, afterpulsing, and crosstalk, which are major concerns when large arrays are being designed.

8.2   CMOS APDs FOR TIME-RESOLVED IMAGING

Unlike in photomultiplier tubes and microchannel plates [42], which are nonsolid-state devices, and electron-multiplied charge-coupled devices [43,44], which have a finite optical gain, in SPADs, the optical gain is virtually infinite, thus enabling in principle the detection of individual photons. Single-photon detection is also possible in ultra-low-noise active pixel sensor architectures [45,46]; however, in these devices, time-resolved imaging is usually limited in resolution due to the relatively low speed of the detection and readout circuits. CCDs can be very fast but still be limited to a microsecond or so [47], while CCD streak cameras can achieve a resolution of a few picoseconds, but they require a two-dimensional pixel array to resolve a string of photon arrivals. Moreover, long acquisition latency and the added complexity to form and deflect the photoelectron beam make the device unsuitable for miniaturization and low-cost operation.

Proportional APDs and SPADs generally have a very low timing response uncertainty [48]. A high timing resolution is achieved, thanks to a fast photomultiplication *in situ* upon the absorption of the seed photon in the p-n junction. Carrier multiplication is achieved through impact ionization, thus producing an avalanche. Both types of APDs can reach timing uncertainties as low as a few tens of picoseconds, thanks to the speed at which an avalanche evolves from the initial carrier pair forming in the multiplication region.

SPADs for time-resolved imaging have evolved over the years, which were aided by novel designs and fabrication process enhancements [49–58]. With the introduction of SPAD devices in CMOS [21] and later in DSM CMOS, the realization of complex circuitry on chip, side by side with the SPAD, has become feasible [59–68]. As the multipixel-based CMOS SPAD imager gets adopted in various applications [33,69–76], the need to enhance the performance of the device is also increasing.

8.3 NOISE PERFORMANCE IN DSM CMOS SPADs

The differentiating factors are essentially the nature and depth of the semiconductor layers where the multiplication region is located and the mechanism that is used to avoid edge breakdown [77–79]. These designs are usually afflicted by a relatively high dark noise, characterized in SPADs by the dark count rate (DCR), mainly due to band-to-band tunneling and trap-assisted tunneling that results from reduced annealing and drive-in diffusion steps [79]. Recently, designs have emerged utilizing lightly doped layers that have helped reduce tunneling noise [30], while, in other designs, the noise was also reduced to 0.05 Hz/μm^2 with the use of special enrichment implants [80].

Though noise was brought down to acceptable levels, for applications such as fluorescence lifetime imaging microscopy (FLIM), Förster resonance energy transfer (FRET), and fluorescence correlation spectroscopy, there is a constant trend toward higher spectral sensitivity. Unfortunately, in standard layers found in CMOS, technologies are not amenable to construct a sensor with wide spectral response. In [81–83], constructions with specific junctions were reported that were either using a substrate as one of the junction nodes or a substrate acting as a photon collection region. In these designs though, the depletion region formed with a substrate requires *ad hoc* isolation mechanisms to isolate the depletion region from transistors and from the adjacent SPADs to reduce crosstalk. Though a sensor designed with this construction was proposed as a silicon photomultiplier, where crosstalk is less problematic, we do not believe that this solution is appropriate in most image sensors [84]. For this reason, hereafter, we will limit our discussion to substrate-isolated SPAD structures.

8.4 CMOS SPADs IN FULL ISOLATION

Let us consider a SPAD that is fabricated in a 180-nm CMOS process; the structure of the device is depicted in the 3D plot of Figure 8.1. The figure shows a circular SPAD with a diameter of 12 μm (active region). In this design, the main junction was engineered to be between p+ and deep n-well-2, referred to as deep n-well—DNW. The DNW, having lower dopant concentration near the junction, helps reduce tunneling noise and enhances spectral response with a wider depletion region, provided that a higher voltage is used in excess of the breakdown. This voltage is known as excess bias voltage, also known as overvoltage; a higher excess bias is sometimes preferable due to the compression of the PDP that in turn causes a reduced sensitivity to breakdown voltage variations. To support a higher excess bias, the guard ring must be optimized to avoid PEB at higher-bias conditions. Buried-N-enabling substrate isolation also provides a contact path to DNW, thus avoiding the need for counter-doping in the guard ring using DNW/n-well as in [5], [79], and [80]. The resulting breakdown voltage of the guard ring is 35.7 V, which is 12.2 V higher than the main junction, as described in the current-to-voltage (I-V) characteristics of Figure 8.2.

In order to maintain a uniform and large sensitive region, it is necessary to maximize the electric field in the center of the active region while suppressing it at the edges. This is achieved in a number of ways; see [85] for a detailed description of various PEB prevention methods. Figure 8.3 shows one such technique based on the use of a guard ring. The figure shows a uniform electric field in the active region, while a quickly tapering field prevents an avalanche formation at the edges. Of course, the drawback of this technique is a loss of sensitivity at the edge of the active area, which is shown in the figure, and it is quantified in Figure 8.4 as a function of wavelength.

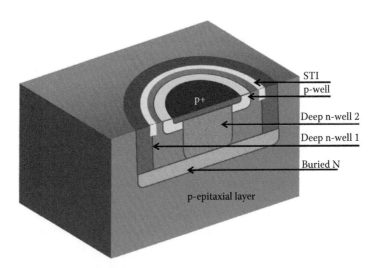

FIGURE 8.1   Cross section of a SPAD fabricated in a 180-nm CMOS technology that was implemented to enable wide-spectral response and isolation from adjacent SPADs and from the surrounding CMOS circuitry. STI, shallow trench isolation.

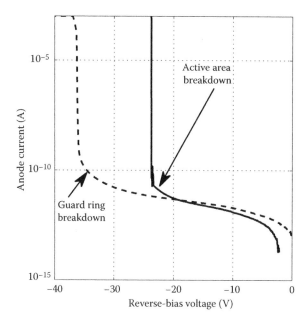

FIGURE 8.2 I-V characteristics of the SPAD, with the breakdown regions shown in the same plot.

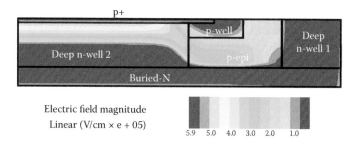

FIGURE 8.3 Techniques for the prevention of PEB in planar processes. The simulation shows how the electric field within the active region is uniform and abruptly decreases at the corners due to the action of the guard ring. This results in an inactive space at the edge of the active region where the PEB is usually located.

In order to operate in Geiger mode, the device must be equipped with a quenching mechanism to stop the avalanche before the temperature in the junction rises above critical values prompting the destruction of the device. Quenching can be active or passive; a discussion of the advantages and disadvantages of various approaches is beyond the scope of this chapter; in what follows, we used a passive quenching scheme that is based on a ballast resistor followed by a comparator. A slow active recharge scheme was used to prevent a lambda-shaped response and to control dead time precisely.

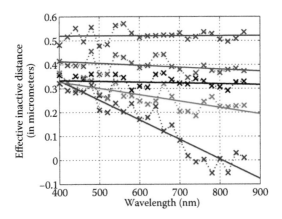

FIGURE 8.4 Inactive space around the active region of a SPAD within the guard ring as a function wavelength.

8.5 TRADING OFF NOISE WITH SENSITIVITY

DCR measurements involve the characterization of a primary and secondary pulse [85]. Primary pulses (or primary DCRs) are random avalanche pulses due to band-to-band tunneling, trap-assisted processes, or a combination of them. Figure 8.5 plots the DCR as a function of excess bias at room temperature for a number of detector batches. One can see the relative stability of DCR from chip to chip and from detector to detector. A much larger dependence is from excess bias and temperature. The temperature dependence of DCR is shown in the Arrhenius plot that is depicted in Figure 8.6. The figure also shows the dependence of DCR from excess bias.

Secondary pulses (also known as afterpulses), on the contrary, are correlated to primary pulses in time and are due to the trapping and detrapping of a carrier that is

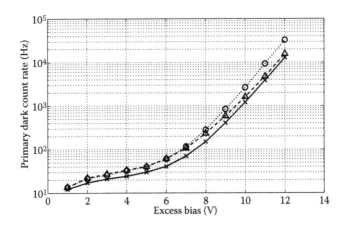

FIGURE 8.5 DCR as a function of excess bias voltage measured at room temperature for various measurement batches.

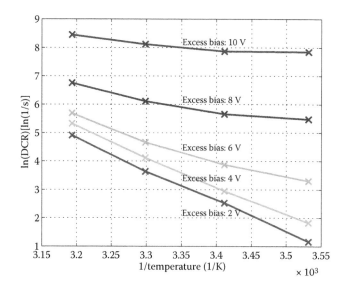

FIGURE 8.6 Arrhenius plot of DCR as a function of temperature for various excess bias voltages. Note the common trend of DCR, irrespective of excess bias and the relative independence of DCR from excess bias at high values of excess bias.

created during the previous avalanche. Secondary pulses were characterized using the interarrival time histogram method, as devised in [86]. In this method, a histogram of all the interarrival times in a SPAD in the dark or in moderate illumination is constructed. This histogram is used to derive the afterpulsing probability as a function of dead time as the integral of the counts in excess of the exponential fit of the histogram. Figure 8.7 shows the interarrival statistics that are measured on the SPAD being reported.

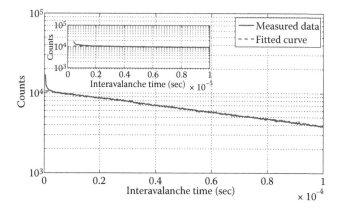

FIGURE 8.7 Interarrival statistics measured on the SPAD object of this chapter. From the resulting interarrival histogram, the probability of afterpulsing is derived as a function of dead time (see text). In the inset, a detail of the histogram is shown for short interarrival times.

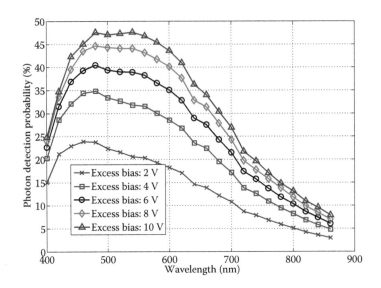

FIGURE 8.8   Photon detection probability as a function of wavelength for a variety of excess bias voltages at room temperature. The PDP becomes increasingly independent of excess bias with its increase, making it possible to achieve high PRNU in very large arrays of SPADs covering large areas of real estate.

Figure 8.8 shows a plot of PDP as a function of these two parameters at room temperature. For a device of 12 μm in active diameter, the peak sensitivity is 47.6% at 480-nm wavelength when biased at 10-V excess bias; the SPAD achieves more than 40% from 440 to 620 nm and more than 30% from 420 to 680 nm at 10-V excess bias. This wide PDP profile is attributed to the wide depletion in DNW and to the device design, facilitating operation at higher excess bias. Furthermore, the rate of increase in PDP tends to decrease at higher excess biases, implying a reduced impact of breakdown and supply voltage variations on PDP, and thus achieving a virtually ideal photo response nonuniformity (PRNU) across the array when realized in imagers.

8.6   DYNAMIC PERFORMANCE

SPADs require a quenching and recharge circuit to prevent their destruction and to enable subsequent photon detections. There exist a number of techniques to achieve this functionality, known as active or passive quenching and recharge. The simplest passive quenching technique is the use of a ballast resistor, which senses the avalanche current and quenches it by bringing the bias voltage of the SPAD from above the breakdown voltage to below it. The ballast and sense resistances can be implemented in polysilicon [87] or using the nonlinear characteristics of a PMOS or an NMOS that is biased in weak or strong inversion [88,89]. In active quenching, the avalanche current is used to actively stop the avalanche. The literature on active quenching is extensive [85,87,90].

During recharge, the photodiode bias voltage must return to the preavalanche state as quickly as possible. Again, there are passive and active schemes to achieve recharge. In passive recharge, the diode will automatically recharge via the ballast resistance. The recharge, in this case, follows the RC exponential, where R is the equivalent quenching resistance and C is the total parasitic capacitance in the diode. In active recharge schemes, the photodiode is forced to the initial state generally via a fast switch that is controlled by a current sense amplifier. Even though these schemes are attractive, they usually require extra complexity to a pixel, thus potentially hindering miniaturization unless an in-pixel feedback is used. In addition, active recharge may result in increased afterpulsing probability if the recharge time is lower than the device's intrinsic relaxation time. Recently, an active recharge scheme was introduced to achieve a different goal, i.e., to fix the recharge time to a predefined value, possibly above intrinsic relaxation, so as to keep afterpulsing within predefined boundaries [88,89].

The quenching and recharge times are collectively known as *dead time*. Dead time determines the dynamic range of SPAD imagers, defined as the ratio of the maximum and minimum detectable photon rate [91]. Dead time in passive quenching/recharge methods is potentially longer than that of their active counterparts. However, the advantage of a reduced dead time in a large array may be decreased by the limited speeds of pixel readout schemes.

8.7   TIMING RESOLUTION

Timing resolution relates to the uncertainty of the response of a SPAD to a short excitation pulse, usually achieved through a laser source. In this measurement, a laser source operated at 40 MHz emitting light pulses within a few picoseconds of timing jitter was used along with the neutral density filters to attenuate light to a single-photon level. Under such experimental conditions, the time difference between the laser output trigger and the SPAD-raising edge was measured using a Lecroy 8600A oscilloscope. The measured statistical distribution of the time difference was used in estimating the device jitter in terms of full width half maximum (FWHM). Jitter measurements performed using two different lasers emitting 405- and 637-nm wavelength light are presented in Figure 8.9. As expected, the red laser causes a longer tail in the response of the SPAD due to the photocarriers that were created in the quasi-neutral region (toward the lower end of the DNW) diffusing toward the depletion region. At higher excess bias, jitter reduces due to the increase in field strength. However, in addition to device jitter, the estimated jitter includes the jitter contributions from the laser (25 ps for blue and 37 ps for red), external circuitry, and the oscilloscope. Using this experimental setup, we were able to obtain a 70-ps FWHM at 10-V excess bias and an 86-ps FWHM using a 405- and a 637-nm laser, respectively.

8.8   TRENDS AND COMPARISONS

The first SPAD array in a 0.35-μm CMOS technology demonstrated fully scalable pixels at a pitch of 25 μm. However, for a realistic pixel sensor realization, this limit should be further reduced. Pixel miniaturization has other benefits too. The

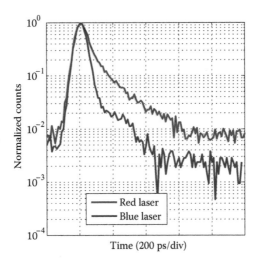

FIGURE 8.9 Time response of the SPAD for two distinct excitation wavelengths. As expected, shorter wavelengths are conducive to better resolutions, thanks to reduced uncertainty or a less prominent tail in the response.

reduction of anode and cathode areas in SPADs generally reduces the DCR. In addition, the number of carriers involved in an avalanche are reduced, thus decreasing the probability of carrier trapping and, consequently, of afterpulsing. Finally, fewer carriers involved in impact ionization will cause smaller photon emission during the avalanche and thus less optical crosstalk.

Furthermore, unlike conventional pixel arrays, where the intensity of light is coded in terms of voltage that is stored in a parasitic capacitance, in SPAD arrays, the state of photon counts is not statically available on pixel. Photon counts can only be stored on pixel when a counter and a memory are integrated *in situ*, with the consequence of increasing the pitch. In addition, while a large memory is beneficial in increasing the counting resolution, and thus the dynamic range for a given readout speed, it further absorbs area and power, thus further increasing the pitch and decreasing the size of a feasible array. An alternative to this approach is the use of a memory of minimum size, 1 bit, or an analog counter. Both techniques have been shown successfully in [76] and [39,63], respectively.

Figure 8.10 plots the PDP of various DSM CMOS SPADs in full isolation, as reported in the literature. Among various substrate-isolated devices, SPADs designed with p+/n-well junctions, as reported by Bronzi [80], Leitner [92], Niclass [79], and Gersbach [78], have resulted in narrower PDP profiles due to the formation of a shallower junction. Although Richardson [30] improved the DCR by design using a p-well/DNW junction, the device spectral response remained identical to conventional designs [78,79,92,93]. A similar design to Richardson's [30], designed in high-voltage CMOS process by Wu [94], has resulted in a wider PDP profile due to the use of lightly doped high-voltage p-well. However, the peak PDP is <25% even at 15-V excess bias. The device presented here attains >40% PDP from 440- to 620-nm wavelength at 10-V excess bias; to the best of our knowledge, the achieved

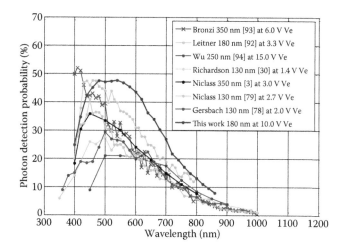

FIGURE 8.10   State of the art of PDP as found in the literature as a function of wavelength and a given excess bias.

performance is superior to any other substrate-isolated CMOS SPAD that was reported in the literature. The high excess bias may pose problems in integrating this design in large arrays unless measures are taken to divide the voltage before applying to digital electronics, as suggested, for example, in [95]. Another alternative is a hybrid approach, where different technologies are combined in a nonmonolithic chip with microbumps and through-silicon vias connecting the circuits [96].

Further, the device's noise performance, compared with the state of the art in Figure 8.11, shows that the noise performance attained in this work is better than most of the CMOS SPADs, and it is comparable with the devices that are presented

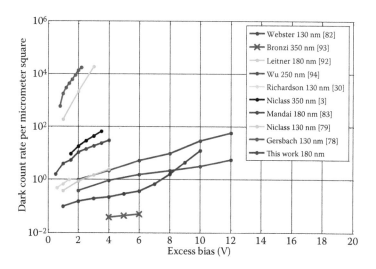

FIGURE 8.11   State of the art of DCR as a function of excess bias at room temperature.

in [94] and [82]. A comprehensive performance analysis of published SPADs is freely available in http://aqua.epfl.ch/spads.

8.9 CONCLUSIONS

In this chapter, we introduced some of the fundamental parameters in proportional APDs and SPADs that are relevant when one wants to build large arrays of photon-counting pixels. We then focused on wide-spectrum SPADs that also have the feature of being isolated from the substrate, thus enabling low crosstalk (both electrical and optical). Thanks to a careful guard ring design-enabling device operation until 10-V excess bias and to a wide depletion region that is achieved using DNW. To the best of our knowledge, the presented device surpasses the performance of all known DSM CMOS SPADs with substrate isolation. The SPAD exhibits a PDP that is extended above 40% in the 440–620-nm spectral range and that is scalable to large arrays of identical pixels. A comparison with the most recent literature is also presented.

REFERENCES

1. E. Charbon. 2014. Single-photon imaging in CMOS processes. *Philosophical Transactions of the Royal Society A*, Vol. 372, No. 2012, doi: 10.1098/rsta.2013.0100.
2. W. Denk, J. H. Stricker, and W. W. Webb. 1990. 2-Photon laser scanning fluorescence microscopy. *Science*, Vol. 248, pp. 73–76.
3. C. Niclass, A. Rochas, P. A. Besse, and E. Charbon. 2004. A CMOS single photon avalanche diode array for 3D imaging. *IEEE International Solid-State Circuits Conference* (ISSCC), pp. 120–121, San Fancisco, February 15–19.
4. C. Niclass and E. Charbon. 2005. A single photon detector array with 64 × 64 resolution and millimetric depth accuracy for 3D imaging. *IEEE International Solid-State Circuits Conference* (ISSCC), pp. 364–365, San Francisco, February 10.
5. C. Niclass, A. Rochas, P. A. Besse, and E. Charbon. 2005. Design and characterization of a CMOS 3-D image sensor based on single photon avalanche diodes. *IEEE Journal of Solid-State Circuits*, Vol. 40, No. 9, pp. 1847–1854, September.
6. D. Stoppa et al. 2007. A CMOS 3-D imager based on single photon avalanche diode. *IEEE Transactions on Circuits and Systems I*, Vol. 54, No. 1, pp. 4–12, January.
7. A. Grinvald et al. 1999. In-vivo optical imaging of cortical architecture and dynamics. In *Modern Techniques in Neuroscience Research*, eds. U. Windhorst and H. Johansson, pp. 893–969, Berlin Heidelberg: Springer.
8. J. Fisher et al. 2004. In vivo fluorescence microscopy of neuronal activity in three dimensions by use of voltage-sensitive dyes. *Optics Letters*, Vol. 29, No. 1, pp. 71–73, January.
9. S. Eisenberg et al. 2002. Visualization and PIV measurements of high-speed flows and other phenomena with novel ultra-high-speed CCD camera. *Proceedings of SPIE*, Vol. 4948, pp. 671–676.
10. S. V. Tipinis et al. 2002. High-speed X-ray imaging camera for time-resolved diffraction studies. *IEEE Transactions on Nuclear Science*, Vol. 49, No. 5, pp. 164–167, October.
11. W. Reckers et al. 2002. Investigation of flame propagation and cyclic combustion variations in a DISI engine using synchronous high-speed visualization and cylinder pressure analysis. *International Symposium für Verbrennungdiagnostik*, pp. 27–32.
12. W. Becker, A. Bergmann, E. Haustein, Z. Petrasek, P. Schwille, C. Biskup, L. Kelbauskas et al. 2006. Fluorescence lifetime images and correlation spectra obtained by multidimensional time-correlated single photon counting. *Microscopy Research and Technique*, Vol. 69, pp. 186–195.

13. W. Becker. 2005. Advanced time-correlated single photon counting techniques. In *Springer Series in Chemical Physics*, eds. Professor A. W. Castleman Jr., Professor J. P. Toennies, Professor W. Zinth. Vol. 81, Berlin, Heidelberg: Springer.

14. J. R. Lakowicz. 1999. *Principles of Fluorescence Spectroscopy*, 2nd edition. Springer-Verlag: Kluwer Academic/Plenum Publishers.

15. R. Hanbury Brown and R. Q. Twiss. 1956. Correlation between photons in two coherent beams of light. *Nature*, Vol. 177, pp. 27–29.

16. R. J. Glauber. 1963. Coherent and incoherent states of the radiation field. *Physical Review*, Vol. 131, pp. 2766–2788.

17. D. L. Boiko, N. J. Gunther, N. Brauer, M. Sergio, C. Niclass, G. B. Beretta, and E. Charbon. 2009. A quantum imager for intensity correlated photons. *New Journal of Physics*, Vol. 11, 013001.

18. R. J. McIntyre. 1985. Recent developments in silicon avalanche photodiodes. *Measurement*, Vol. 3, No. 4, pp. 146–152.

19. S. Cova, A. Longoni, and A. Andreoni. 1981. Towards picosecond resolution with single-photon avalanche diodes. *Review of Scientific Instruments,* Vol. 52, No. 3, pp. 408–412.

20. W. J. Kindt. 1999. *Geiger Mode Avalanche Photodiode Arrays for Spatially Resolved Single Photon Counting*. PhD Thesis, Delft, Netherlands: Delft University Press, ISBN 90-407-1845-8.

21. A. Rochas, M. Gani, B. Furrer, P.-A. Besse, R. Popovic, G. Ribordy, and N. Gisin. 2003. Single photon detector fabricated in a complementary metal oxide semiconductor high-voltage technology. *Review of Scientific Instruments*, Vol. 74, No. 7, pp. 3263–3270.

22. C. Niclass, M. Sergio, and E. Charbon. 2006a. A 64 × 48 single photon avalanche diode array with event-driven readout. *European Solid-State Circuits Conference* (ESSCIRC), pp. 556–559, Montreux, September.

23. C. Niclass, M. Sergio, and E. Charbon. 2006b. A single photon avalanche diode array fabricated in deep-submicron CMOS technology. *Design Automation & Test in Europe* (DATE), pp. 1–6, Munich, March 6–10.

24. C. Niclass, M. Sergio, and E. Charbon. 2006c. A single photon avalanche diode array fabricated in 0.35 μm CMOS and based on an event-driven readout for TCSPC experiments. In *Advanced Photon Counting Techniques Meeting* (Boston), ed. W. Becker, *Proc. SPIE* 6372 paper 637205.

25. D. Mosconi et al. 2006. CMOS single-photon avalanche diode array for time-resolved fluorescence detection. *European Solid-State Circuits Conference* (ESSCIRC), pp. 564–567, Montreux, September.

26. B. Rae et al. 2008. A microsystem for time-resolved fluorescence analysis using CMOS single-photon avalanche diodes and Micro-LEDs. *IEEE International Solid-State Circuits Conference*, pp. 166–603, San Francisco, February 3–7.

27. S. Tisa, F. Guerrieri, A. Tosi, and F. Zappa. 2008. 100kframe/s 8 bit monolithic single-photon imagers. *European Solid-State Device Research Conference*, pp. 274–277, Edinburgh, September 15–19.

28. M. A. Marwick and A. G. Andreou. 2008. Single photon avalanche photodetector with integrated quenching fabricated in TSMC 0.18 μm CMOS process. *Electronics Letters*, Vol. 44, No. 10, 643–644, October.

29. N. Faramarzpour, M. J. Deen, S. Shirani, and Q. Fang. 2008. Fully integrated single photon avalanche diode detector in standard CMOS 0.18-μm technology. *IEEE Transactions on Electron Devices*, Vol. 55. No. 3, pp. 760–767, March.

30. J. Richardson, L. Grant, and R. Henderson. 2009. Low dark count single-photon avalanche diode structure compatible with standard nanometer scale CMOS technology. *IEEE Photonics Technology Letters*, Vol. 21, No. 14, pp. 1020–1022.

31. M. Gersbach, C. Niclass, J. Richardson, R. Henderson, L. Grant, and E. Charbon. 2008. A single-photon detector implemented in a 130 nm CMOS imaging process. *European Solid-State Device Research Conference*, pp. 274–277, Edinburgh, September 15–19.

32. M. Gersbach, J. Richardson, E. Mazaleyrat, S. Hardillier, C. Niclass, R. Henderson, L. Grant, and E. Charbon. 2009. A low-noise single-photon detector implemented in a 130 nm CMOS imaging process. *Solid-State Electronics*, Vol. 53, No. 7, pp. 803–808, July.

33. M. Sergio, C. Niclass, and E. Charbon. 2007. A 128 × 2 CMOS single-photon streak camera with timing-preserving latchless pipeline readout. *IEEE International Solid-State Circuits Conference* (ISSCC), pp. 394–610, San Francisco, February 11–15.

34. R. Walker, J. Richardson, and R. Henderson. 2011. A 128 × 96 pixel event-driven phase-domain $\Sigma\Delta$-based fully digital 3D camera in 0.13 µm CMOS imaging technology. *IEEE International Solid-State Circuits Conference* (ISSCC), pp. 410–412, San Francisco, February 20–24.

35. C. Niclass, C. Favi, T. H. Kluter, M. Gersbach, and E. Charbon. 2008. A 128 × 128 single-photon imager with on-chip column-level 10b time-to-digital-converter array capable of 97ps resolution. *IEEE International Solid-State Circuits Conference* (ISSCC), pp. 44–45, San Francisco, February 3–7.

36. L. Carrara, C. Niclass, N. Scheidegger, H. Shea, and E. Charbon. 2009. A gamma, X-ray and high energy proton radiation-tolerant CIS for space applications. *IEEE International Solid-State Circuits Conference* (ISSCC), pp. 40–41, San Francisco, February 8–12.

37. S. Burri, Y. Maruyama, X. Michalet, F. Regazzoni, C. Bruschini, and E. Charbon. 2014. Architecture and applications of a high resolution gated SPAD image sensor. *Optics Express*, Vol. 22, No. 14, pp. 17573–17589, July.

38. O. Shcherbakova, L. Pancheri, G.-F. Dalla Betta, N. Massari, and D. Stoppa. 2013. 3D camera based on linear-mode gain-modulated avalanche photodiodes. *IEEE International Solid-State Circuits Conference* (ISSCC), San Francisco, February 17–21.

39. N. Dutton, L. Parmesan, A. Holmes, L. Grant, and R. K. Henderson. 2014. 320√6240 oversampled digital single photon counting image sensor. *Symposium on VLSI Circuits*, June.

40. E. Charbon, L. Carrara, C. Niclass, N. Scheidegger, and H. Shea. 2010. Radiation-tolerant CMOS single-photon imagers for multi-radiation detection. In *Radiation Effects in Semiconductors: Devices, Circuits, and Systems*, pp. 31–50, ed. K. Iniewski.

41. E. Charbon. 2004. Will CMOS imagers ever need ultra-high speed? *IEEE International Conference on Solid-State and IC Technology*, Vol. 3, pp. 1975–1980, Beijing, October 18–21.

42. J. McPhate, J. Vallerga, A. Tremsin, O. Siegmund, B. Mikulec, and A. Clark. 2004. Noiseless kilohertz-frame-rate imaging detector based on microchannel plates readout with Medipix2 CMOS pixel chip. *Proc. SPIE*, Vol. 5881, pp. 88–97.

43. T. G. Etoh et al. 2005. Design of the PC-ISIS: Photon-counting *in-situ* storage image sensor. *IEEE Workshop on CCDs and Advanced Image Sensors*, pp. 113–116, Karuizawa, Japan, June 9–11.

44. J. Hynecek. 2001. Impactron—A new solid state image intensifier. *IEEE Transactions on Electron Devices*, Vol. 48, No. 10, pp. 2238–2241, October.

45. N. Kawai and S. Kawahito. 2005. A low-noise signal readout circuit using double-stage noise cancelling architecture for CMOS image sensors. *IEEE Workshop on CCDs and Advanced Image Sensors*, pp. 27–30, Karuizawa, Japan, June 9–11.

46. R. Shimitzu et al. 2009. A charge-multiplication CMOS image sensor suitable for low-light-level imaging. *IEEE International Solid-State Circuits Conference* (ISSCC), pp. 50–51, San Francisco, February 8–12.

47. T. G. Etoh et al. 2003. An image sensor which captures 100 consecutive frames at 1,000,000 frames/s. *IEEE Transactions on Electron Devices*, pp. 144–151, Vol. 50, No. 1, pp. 144–151, January.

48. R. H. Haitz. 1965. Studies on optical coupling between silicon p-n junctions. *Solid-State Electronics*, Vol. 8, pp. 417–425.
49. M. Ghioni, S. Cova, A. Lacaita, and G. Ripamonti. 1988. New silicon epitaxial avalanche diode for single-photon timing at room temperature. *Electronics Letters*, Vol. 24, No. 24, pp. 1476–1477.
50. A. Lacaita, M. Ghioni, and S. Cova. 1989. Double epitaxy improves single-photon avalanche diode performance. *Electronics Letters*, Vol. 25, No. 13, pp. 841–843.
51. S. Cova, A. Lacaita, M. Ghioni, G. Ripamonti, and T. A. Louis. 1989. 20ps timing resolution with single photon avalanche diodes. *Review of Scientific Instruments*, Vol. 60, No. 6, pp. 1104–1110.
52. M. Ghioni, A. Gulinatti, P. Maccagnani, I. Rech, and S. Cova. 2006. Planar silicon SPADs with 200 μm diameter and 35 ps photon timing resolution. In Optics East 2006 (pp. 63720R–63720R). International Society for Optics and Photonics, October.
53. A. Lacaita, S. Cova, M. Ghioni, and F. Zappa. 1993. Single-photon avalanche diode with ultrafast pulse response free from slow tails. *IEEE Electron Device Letters*, Vol. 14, No. 7, pp. 360–362.
54. B. Aull, A. Loomis, J. A. Gregory, and D. Young. 1998. Geiger-mode avalanche photodiode arrays integrated with CMOS timing circuits. *56th Annual Device Research Conference Digest*, pp. 58–59, Charlottesville, VA, June 22–24.
55. J. C. Jackson, A. Morrison, D. Phelan, and A. Mathewson. 2002. A novel silicon Geiger-mode avalanche photodiode. *IEEE International Electron Device Meeting* (IEDM), pp. 797–800, San Francisco, December 8–11.
56. E. Sciacca, A. Giudice, D. Sanfilippo, F. Zappa, S. Lombardo, R. Cosentino, C. Di Franco et al. 2003. Silicon planar technology for single-photon optical detectors. *Transactions on Electron Devices*, Vol. 50, No. 4, pp. 918–925.
57. M. Ghioni, G. Armellini, P. Maccagnani, I. Rech, M. Emsley, and M. Unlu. 2008. Resonant-cavity-enhanced single-photon avalanche diodes on reflecting silicon substrates. *IEEE Photonics Technology Letters*, Vol. 20, No. 6, pp. 413–415.
58. B. Aull, J. Burns, C. Chen, B. Felton, H. Hanson, C. Keast, J. Knecht et al. 2006. Laser radar imager based on 3D integration of Geiger-mode avalanche photodiodes with two soi timing circuit layers. *IEEE International Solid-State Circuits Conference* (ISSCC), pp. 1179–1188, San Francisco, February 6–9.
59. S. Tisa, A. Tosi, and F. Zappa. 2007 Fully-integrated CMOS single photon counter. *Optics Express*, Vol. 15, No. 6, pp. 2873–2887, March.
60. A. Rochas, M. Gosch, A. Serov, P. A. Besse, R. Popovic, T. Lasser, and R. Rigler. 2003. First fully integrated 2-D array of single-photon detectors in standard CMOS technology. *IEEE Photonics Technology Letters*, Vol. 15, No. 7, pp. 963–965.
61. F. Zappa, A. Gulinatti, P. Maccagnani, S. Tisa, and S. Cova. 2005. SPADA: Single-photon avalanche diode arrays. *IEEE Photonics Technology Letters*, Vol. 17, No. 3, pp. 657–659.
62. C. Veerappan, J. Richardson, R. Walker, D.-U. Li, M. Fishburn, Y. Maruyama, D. Stoppa et al. 2011. A 160×128 single-photon image sensor with on-pixel 55ps 10b time-to-digital converter. *IEEE Solid-State Circuits Conference* (ISSCC), pp. 312–314, San Francisco, February 20–24.
63. D. Stoppa, F. Borghetti, J. Richardson, R. Walker, L. Grant, R. Henderson, M. Gersbach, and E. Charbon. 2009. A 32×32-pixel array with in-pixel photon counting and arrival time measurement in the analog domain. *European Solid-State Circuits Conference* (ESSCIRC), pp. 204–207, Athens, September 14–18.
64. M. Gersbach, Y. Maruyama, E. Labonne, J. Richardson, R. Walker, L. Grant, R. Henderson, F. Borghetti, D. Stoppa, and E. Charbon. 2009. A parallel 32 × 32 time-to-digital converter array fabricated in a 130 nm imaging CMOS technology. *European Solid-State Circuits Conference* (ESSCIRC), pp. 196–199, Athens, September 14–18.

65. J. Richardson, R. Walker, L. Grant, D. Stoppa, F. Borghetti, E. Charbon, M. Gersbach, and R. Henderson. 2009. A 32 × 32 50ps resolution 10bit time-to-digital converter array in 130 nm CMOS for time correlated imaging. *IEEE Custom Integrated Circuits Conference* (CICC), pp. 77–80, San Jose, September 13–16.

66. S. Mandai and E. Charbon. 2013. Timing optimization of a H-tree based digital silicon photomultiplier. *Journal of Instrumentation*, Vol. 8, No. 9, pp. P09016.

67. D. Tyndall, B. Rae, D. Li, J. Richardson, J. Arlt, and R. Henderson. 2012. A 100mphoton/s time-resolved mini-silicon photomultiplier with on-chip fluorescence lifetime estimation in 0.13 μm CMOS imaging technology. *IEEE International Solid-State Circuits Conference* (ISSCC), pp. 122–124, February.

68. F. Villa, B. Markovic, S. Bellisai, D. Bronzi, A. Tosi, F. Zappa, S. Tisa et al. 2012. SPAD smart pixel for time-of-flight and time-correlated single-photon counting measurements. *IEEE Photonics Journal*, Vol. 4, No. 3, pp. 795–804, June.

69. D. D.-U. Li, J. Arlt, D. Tyndall, R. Walker, J. Richardson, D. Stoppa, E. Charbon, and R. K. Henderson. 2011. Video-rate fluorescence lifetime imaging camera with CMOS single-photon avalanche diode arrays and high-speed imaging algorithm. *Journal of Biomedical Optics*, Vol. 16, No. 9, pp. 096 012–096 012–12.

70. D. Mosconi, D. Stoppa, L. Pancheri, L. Gonzo, and A. Simoni. 2006. CMOS single-photon avalanche diode array for time-resolved fluorescence detection. *European Solid-State Circuits Conference* (ESSCIRC), pp. 564–567, Montreux, September.

71. Y. Maruyama, J. Blacksberg, and E. Charbon. 2013. A 1024×8 700ps time-gated SPAD line sensor for laser Raman spectroscopy and LIBS in space and rover-based planetary exploration. *IEEE Solid-State Circuits Conference Digest* (ISSCC), pp. 110–111, February.

72. C. Niclass and E. Charbon. 2005. A single photon detector array with 64 × 64 resolution and millimetric depth accuracy for 3D imaging. *IEEE International Solid-State Circuits Conference* (ISSCC), pp. 364–604, San Francisco, February 10.

73. D. Stoppa, L. Pancheri, M. Scandiuzzo, L. Gonzo, G.-F. Dalla Betta, and A. Simoni. 2007. A CMOS 3-D imager based on single photon avalanche diode. *IEEE Transactions on Circuits and Systems I*: Regular Papers, Vol. 54, No. 1, pp. 4–12, January.

74. T. Frach, G. Prescher, C. Degenhardt, R. de Gruyter, A. Schmitz, and R. Ballizany. 2009. The digital silicon photomultiplier—Principle of operation and intrinsic detector performance. *IEEE Nuclear Science Symposium* (NSS/MIC), pp. 1959–1965, Orlando, October 24–November 1, 2009.

75. L. H. Braga, L. Gasparini, L. Grant, R. Henderson, N. Massari, M. Perenzoni, D. Stoppa, and R. Walker. 2013. An 8×16-pixel 92kSPAD time-resolved sensor with on-pixel 64ps 12b TDC and 100ms/s real-time energy histogramming in 0.13 μm CIS technology for PET/MRI applications. *IEEE International Solid-State Circuits Conference* (ISSCC), pp. 486–487, San Francisco, February 17–21.

76. S. Burri, D. Stucki, Y. Maruyama, C. Bruschini, E. Charbon, and F. Regazzoni. 2013. Two-dimensional mapping of photon counts in low-noise single-photon avalanche diodes. *International Imaging Sensor Workshop* (IISW), Snow Bird, June 12–16.

77. H. Finkelstein, M. Hsu, and S. Esener. 2006. STI-bounded single-photon avalanche diode in a deep-submicrometer CMOS technology. *IEEE Electron Device Letters*, Vol. 27, No. 11, pp. 887–889.

78. M. Gersbach, C. Niclass, E. Charbon, J. Richardson, R. Henderson, and L. Grant. 2008. A single photon detector implemented in a 130 nm CMOS imaging process. *European Solid-State Device Research Conference* (ESSDERC), pp. 270–273, Edinburgh, September 15–19.

79. C. Niclass, M. Gersbach, R. Henderson, L. Grant, and E. Charbon. 2007. A single photon avalanche diode implemented in 130-nm CMOS technology. *IEEE Journal of Selected Topics in Quantum Electronics*, Vol. 13, No. 4, pp. 863–869, August/September.

80. D. Bronzi, F. Villa, S. Bellisai, B. Markovic, S. Tisa, A. Tosi, F. Zappa, S. Weyers, D. Durini, W. Brockherde, and U. Paschen. 2012. Low-noise and large-area CMOS spads with timing response free from slow tails. *European Solid-State Device Research Conference* (ESSDERC), pp. 230–233, Bordeaux, September 17–21.
81. E. Webster, J. Richardson, L. Grant, D. Renshaw, and R. Henderson. 2012. A single-photon avalanche diode in 90-nm CMOS imaging technology with 44 photon detection efficiency at 690 nm. *IEEE Electron Device Letters*, Vol. 33, No. 5, pp. 694–696.
82. E. Webster, L. Grant, and R. Henderson. 2012. A high-performance single-photon avalanche diode in 130-nm CMOS imaging technology. *IEEE Electron Device Letters*, Vol. 33, No. 11, pp. 1589–1591.
83. S. Mandai, M. W. Fishburn, Y. Maruyama, and E. Charbon. 2012. A wide spectral range single-photon avalanche diode fabricated in an advanced 180 nm CMOS technology. *Optics Express*, Vol. 20, No. 6, pp. 5849–5857, March.
84. E. Webster, R. Walker, R. Henderson, and L. Grant. 2012. A silicon photomultiplier with 30% detection efficiency from 450–750 nm and 11.6 μm pitch NMOS-only pixel with 21.6% fill factor in 130 nm CMOS. *European Solid-State Device Research Conference* (ESSDERC), pp. 238–241, Bordeaux, September 17–21.
85. S. Cova, M. Ghioni, A. Lacaita, C. Samori, and F. Zappa. 1996. Avalanche photodiodes quenching circuits for single-photon detection. *Applied Optics*, Vol. 35, No. 12, pp. 1956–1976.
86. M. Fishburn. 2012. *Fundamentals of CMOS Single-Photon Avalanche Diodes*. PhD Dissertation, Delft, Netherlands: Delft University of Technology.
87. A. Rochas. 2003. *Single Photon Avalanche Diodes in CMOS Technology*. PhD Thesis, Lausanne, École Polytechnique Fédérale de Lausanne.
88. C. Niclass, C. Favi, T. Kluter, F. Monnier, and E. Charbon. 2008. Single-photon synchronous detection. *European Solid-State Circuits Conference* (ESSCIRC), pp. 114–117, Edinburgh, September 15–19.
89. C. Niclass, C. Favi, T. Kluter, F. Monnier, and E. Charbon. 2009. Single-photon synchronous detection. *IEEE Journal of Solid-State Circuits*, Vol. 44, No. 7, pp. 1977–1989, July.
90. J. Richardson, R. Henderson, and D. Renshaw. 2007. Dynamic quenching for single photon avalanche diode arrays. *International Imaging Sensor Workshop* (IISW), Ogunquit, June 6–10.
91. A. Eisele, R. K. Henderson, B. Schmidtke, T. Funk, L. Grant, J. Richardson, and W. Freude. 2011. 185 MHz count rate, 139 dB dynamic range single-photon avalanche diode with active quenching circuit in 130 nm CMOS technology. *International Imaging Sensor Workshop* (IISW), Hokkaido, June 8–11.
92. T. Leitner, A. Feiningstein, R. Turchetta, R. Coath, S. Chick, G. Visokolov, V. Savuskan et al. 2013. Measurements and simulations of low dark count rate single photon avalanche diode device in a low voltage 180-nm CMOS image sensor technology. *IEEE Transactions on Electron Devices*, Vol. 60, No. 6, pp. 1982–1988.
93. D. Bronzi, F. Villa, S. Bellisai, S. Tisa, A. Tosi, G. Ripamonti, F. Zappa et al. 2013. Large-area CMOS SPADs with very low dark counting rate. *Proceedings of SPIE* 8631, pp. 86 311B–86 311B–8, February.
94. J.-Y. Wu, S.-C. Li, F.-Z. Hsu, and S.-D. Lin. 2013. Two-dimensional mapping of photon counts in low-noise single-photon avalanche diodes. *International Imaging Sensor Workshop* (IISW), Utah, June 12–16.
95. Z. Xiao, D. Pantic, and R. Popovic. 2007. A new single photon avalanche diode in CMOS high-voltage technology. *TRANSDUCERS*, pp. 1365–1368, Lyon, June 10–14.
96. B. Aull, H. Andrew, J. Douglas, B. J. Richard, M. H. J. Peter, and J. L. Deborah. 2002. Geiger-mode avalanche photodiodes for three-dimensional imaging. *Lincoln Laboratory Journal*, Vol. 13, No. 2, pp. 335–350.

9 Front-End Electronics for Silicon Photomultipliers

Cristoforo Marzocca, Fabio Ciciriello,
Francesco Corsi, Francesco Licciulli,
and Gianvito Matarrese

CONTENTS

9.1 Introduction 203
9.2 SiPM Structure and Electric Model..204
9.3 Front-End Electronics for SiPMs..211
    9.3.1 Charge-Sensitive Amplifier Readout...212
    9.3.2 Voltage-Mode Readout 217
    9.3.3 Current-Mode Readout...222
9.4 Architecture Examples 225
References..234

9.1 INTRODUCTION

Nowadays, detection of low-light levels is required in several application fields, including high-energy and astroparticle physics, medical imaging, and material science, just to mention a few. Calorimetry and positron emission tomography (PET) are among the most common examples of these kinds of applications, which call for the availability of detectors featuring some form of internal charge multiplication mechanism, especially when single-photon sensitivity is required. In the past, photomultiplier tubes (PMTs) and avalanche photodiodes (APDs) have been the detectors of choice to fulfill such requirements, thanks to their excellent characteristics in terms of gain and time resolution. The advent of single-photon avalanche photodiodes (SPADs) operated in Geiger mode paved the way to the introduction of silicon photomultipliers (SiPMs), which offer several advantages over the competitor detectors [1–5]. Low-voltage operation, compactness, ruggedness, and low cost make SiPMs an interesting solution in a broad range of applications and experimental environments. Availability in different pixel numbers and sizes and easy customization guarantee SiPMs' high flexibility and adaptability to a large variety of situations. Furthermore, insensitivity to magnetic fields extends their possible use to frontier applications, such as positron emission tomography–magnetic resonance imaging (PET-MRI) instrumentation. Thanks to the growing interest in, and the diffusion of, these kinds of detectors, increasing research efforts have been dedicated to the development of the related technology, thus allowing, on the one hand, further improvements of the performance in terms of gain, photon detection efficiency, and timing resolution and, on the other hand, mitigation of the main drawbacks that characterize the behavior of SiPMs [6].

Full exploitation of the favorable characteristics of SiPM detectors calls for the availability of suitable front-end electronics. Classic integrated front-end architectures usually employed to read out multichannel radiation detectors cannot be successfully applied or exhibit poor performance in the case of SIPMs because of the peculiar features of these kinds of detectors in terms of equivalent capacitance and large gain. When energy measurements are of interest, very often, large dynamic range requirements must be fulfilled, and simple solutions based on direct charge integration cannot be employed, especially when the allowed voltage headroom is very limited, as happens when deep-submicron standard CMOS technologies, characterized by low supply voltages, are used. In several modern applications, such as time-of-flight positron emission tomography (ToF-PET) [7] or various measurement techniques requiring single-photon detection [8], the most relevant performance to be achieved is timing accuracy, which must be optimized considering the large detector capacitance and the influence of unavoidable parasitic effects. In this case, the architecture and main parameters of front-end electronics play a fundamental role in determining the attainable timing accuracy and must be carefully selected.

To study the behavior and the performance of interest for the system composed by the SiPM and the front-end electronics, an accurate electric model of the detector is needed. Thus, first of all, a discussion about the SiPM models proposed in the literature is mandatory in order to select the most suitable and simple model to perform reliable simulations at circuit level and to gain insight about the relationship between the performance of the system and the parameters of the electronics.

Different front-end architectures and approaches have been adopted in the circuit realizations that are presented in the literature. Exploiting the electric model of the SiPM, the advantages and drawbacks of the proposed solutions will be analyzed so as to identify the most suitable solution, depending on the requirements of the application.

Finally, since, in most applications, multichannel detectors are required, the organization and management of the information provided by each channel within the detection system play a key role to attain the desired performance. Examples of application-specific integrated circuit (ASIC) architectures that adopt interesting solutions will be discussed to give an idea about the range of the available options, the related complexity, and the target applications.

9.2 SiPM STRUCTURE AND ELECTRIC MODEL

As is well known, an SiPM consists of an array of solid-state avalanche photodiodes that are biased above the breakdown voltage V_{BR} and is operated in Geiger mode [1–5]. All the pixels, or microcells, are connected in parallel and biased at the same supply voltage V_{BIAS}, distributed by means of a metal grid that extends all over the surface of the detector, as schematically illustrated in Figure 9.1.

Figure 9.2 shows the working principle of the single microcell. A photon absorbed in the π region creates an electron–hole pair (Figure 9.2a). The electron drifts toward the high field p-n$^+$ region and triggers an avalanche breakdown, which

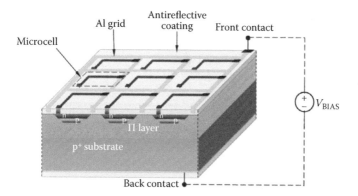

FIGURE 9.1   Structure of the SiPM.

cannot be sustained indefinitely but is quenched by means of a large series resistor R_q (Figure 9.2b). In practice, most of the large amount of charge generated by the very quick phenomenon of the avalanche is prevented from leaving the space charge region of the photodiode through R_q, which carries a negligible current. This charge crowds within the region in which avalanche multiplication takes place and causes a fast decrease in the local electric field: when the total decreasing voltage across the photodiode reaches the breakdown value V_{BR}, the avalanche is quenched, and the current that flows through the quenching resistor slowly brings the structure back to the initial status, carrying the excess charge outside the depletion region, during the so-called recovery phase (Figure 9.2c).

Considering the voltage variation across the photodiode $V_{BIAS} - V_{BR}$, i.e., the overvoltage ΔV, and its equivalent capacitance C_{pixel}, the total amount of charge Q_{TOT} collected as a consequence of an avalanche breakdown triggered in a microcell is given by

$$Q_{TOT} = C_{pixel}(V_{BIAS} - V_{BR}) = C_{pixel}\Delta V. \qquad (9.1)$$

Let us consider the SiPM that is directly connected to an ideal voltage source without any series resistor, as shown in Figure 9.1. In this case, based on the previous discussion, the peak value of the current pulse, which should be observed in response to an event that triggers the avalanche breakdown in one of the microcells, is

$$I_{peak} = \frac{V_{BIAS} - V_{BR}}{R_q}, \qquad (9.2)$$

where R_q is the quenching resistor.

Note that Equation 9.2 holds true under the hypothesis that the avalanche can be considered much faster than the duration of the recovery phase, which is a fully reasonable assumption. However, I_{peak} values larger than the one foreseen by

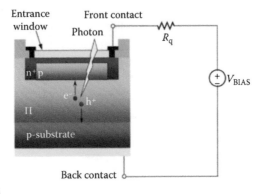

(a)

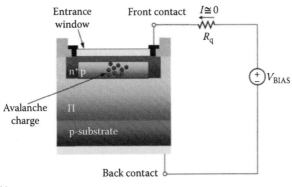

(b)

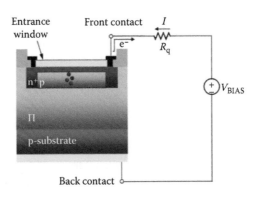

(c)

FIGURE 9.2 Operating principle of the single microcell of the SiPM: (a) photon absorption, (b) avalanche breakdown and quenching, and (c) recovery.

Equation 9.2 are observed in the experiments, when the current pulse of the detector is read out, for instance, by means of the classic circuit that is depicted in Figure 9.3.

Here, the current pulse of the detector is converted into a voltage pulse by means of a series resistor R_{IN}, and a voltage amplifier is used to obtain the desired signal level.

The larger value of the peak current is explained by the existence of a small parasitic capacitance C_q placed across the quenching resistor. C_q represents an alternative fast path for the avalanche current with respect to R_q so that a fraction of the generated charge is able to leave the depletion region of the photodiode through C_q, during the avalanche phenomenon before the quenching happens. Thus, this charge is almost immediately collected by the readout circuit and significantly contributes to the I_{peak} value.

After this short introduction about the behavior and role of the different components of the single microcells of the SiPM in the formation of the output pulse, let us consider the electric model for this detector. Of course, the complete electric model of the SiPM must be based upon the model of the single microcell, i.e., the SPAD.

One of the first electric models for the SPAD has been proposed by Cova et al. [9], and, among its other interesting features, it tries to address the modeling of the avalanche current waveform. In particular, a widespread version of this model [10,11] is illustrated in Figure 9.4, where C_d is the photodiode capacitance.

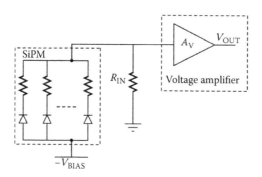

FIGURE 9.3 SiPM readout by means of a series resistor R_{IN} and a voltage amplifier.

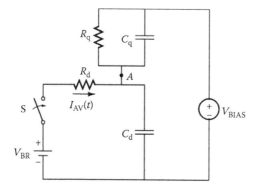

FIGURE 9.4 Model of the SPAD proposed by Cova et al.

Here, the triggering of the avalanche is simulated by closing the switch S, which causes a sudden variation of the current in the resistor R_d from zero to the value $\Delta V/R_d$. Since R_d is in the order of kiloohms, i.e., orders of magnitude smaller than the quenching resistor (hundreds of kiloohms), this corresponds to a very fast variation of the voltage at the internal node A from the initial value V_{BIAS} toward the asymptotic value V_{BR}. As explained above in this section, where the capacitance C_q has been introduced, C_q acts as a fast path for the avalanche current toward the external circuit, and the charge $C_q\Delta V$ is collected very quickly, with a transient behavior that is ruled by the same fast time constant $\tau_d = R_d(C_d + C_q)$, which characterizes the avalanche current $I_{AV}(t)$:

$$I_{AV}(t) = I_{R_d}(t) = \frac{\Delta V}{R_q} e^{-\frac{t}{\tau_d}}. \tag{9.3}$$

Note that the total pixel capacitance $C_{pixel} = C_d + C_q$ is in the order of tens of femtofarads; thus, the order of magnitude of τ_d is tens of picoseconds, corresponding to a very fast transient.

In the model, the quenching of the avalanche is associated to the decrease of the avalanche current, i.e., the current in R_d, and to the behavior of the voltage across the photodetector, which tends to V_{BR}. The start of the recovery phase is simulated by opening the switch when the avalanche current becomes smaller than a reference value I_q (less than 100 μA), under which the avalanche cannot self-sustain. After the opening of the switch, the recovery takes place: the voltage at the internal node A is restored to the initial value V_{BIAS} with an exponential transient that is characterized by the recovery time constant $\tau_r = R_q(C_d + C_q)$, approximately three orders of magnitude larger than τ_d. During the recovery phase, the charge $C_d\Delta V$ is collected by the external circuit, flowing through the parallel $R_q - C_q$, whereas the charge $C_q\Delta V$ circulates within the quenching resistor so that the total amount of collected charge is

$$Q_{TOT} = (C_d + C_q)\Delta V. \tag{9.4}$$

In the following, some drawbacks of the model described above are discussed. First of all, some of the parameters involved are difficult to be extracted from measurements, so their values can only be approximately estimated. In fact, whereas consolidated parameter extraction procedures exist for the capacitances, the quenching resistance, and the breakdown voltage, no similar techniques are available for the resistance R_d and the current I_q. This means that it is very difficult to model with accuracy the behavior of the avalanche current. On the other hand, the simulation of the triggering and quenching of the avalanche by means of the switch is practical only if just one SPAD must be simulated but becomes tricky when several SPAD microcells must be simulated all together within the structure of an SiPM, especially when the incident photons trigger avalanche breakdowns in different microcells at different times, as happens, for instance, when a scintillator is used to detect high-energy gamma photons. In this case, considering a Simulation Program with Integrated Circuit Emphasis (SPICE) implementation of the model, each single switch should be driven by its own control voltage, which, in turn, should depend on the level of the current in R_d, in order to open the switch when this current drops to the value I_q. This means that a rather complex circuit, made up with driven sources, must be set up and replicated for the number of microcells that are contained in the SiPM, which can be very high, also tens of thousands. This makes the described model difficult to be used in several practical cases.

In the vast majority of the applications, it is useless to be able to reproduce with accuracy the transient behavior of the avalanche current because either front-end electronics with the required bandwidth is not available or other time constants introduced by the readout circuit limit the maximum slope of the signal that is produced by the detector. Starting from this practical consideration, a remarkable simplification of the model in Figure 9.4 can be obtained if we model the avalanche current simply by means of a current source that delivers the total amount of charge that is collected when an avalanche breakdown is triggered in a time that is much shorter than the time constants that are introduced by the limited bandwidth of the front-end electronics or, in any case, by the external circuit [12,13]. In practice, the avalanche current can be modeled by means of a sort of Dirac's delta $Q_{TOT}\delta(t)$, as depicted in Figure 9.5: a simpler circuit result, which is not able to reproduce the waveform of the avalanche current but is useful in almost every practical situation.

Once again, if the SPAD is directly connected to the source of the bias voltage, the charge fraction $C_q \Delta V$ is collected immediately by the external circuit, whereas the rest of the total charge Q_{TOT}, i.e., $C_d \Delta V$, is released according to the recovery time constant τ_r.

As mentioned, the main advantage of the described modification is that the model of the complete SiPM becomes very simple without losing accuracy in practical applications. Figure 9.6 illustrates this point for a detector that is composed by N microcells.

Note that, thanks to the superposition principle, all the current sources that model the avalanche current within each of the microcells can be reduced to only one current source $I_{AV}(t)$, which generates all the Dirac's delta that is associated to each avalanche breakdown that is triggered by M-absorbed photons, according to their arrival times t_i, $i = 1,2,\ldots,M$:

$$I_{AV}(t) = Q_{TOT} \sum_{i=1}^{M} \delta(t - t_i). \tag{9.5}$$

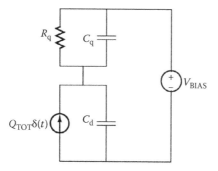

FIGURE 9.5 Simplified model of the SPAD.

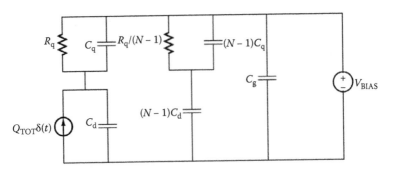

FIGURE 9.6   Complete model of the SiPM.

The parasitic capacitance associated to the large metal grid used to connect in parallel all the microcells and to distribute the bias voltage is modeled in Figure 9.6 by the capacitance C_g, which significantly affects the waveform of the current pulses that are generated by the detector. As already pointed out, robust procedures are available for the extraction of the values of the parameters that are involved in the model of Figure 9.6 from suitable measurements [13,14]. Using these procedures, it is possible to validate the effectiveness of the model in reproducing the waveforms that are generated by an SiPM, coupled to front-end electronics with known characteristics. For example, let us consider a typical SiPM with 3600 microcells 50 × 50 μm from Hamamatsu (S10931-050P). The extraction procedure yields the parameter values that are reported in Table 9.1.

The detector has been read out according to the basic principle that is illustrated in Figure 9.3: two noninverting cascaded operational amplifier (OP AMP) stages, both based upon an LMH6703 OP AMP, have been used to realize the gain stage, which features a gain of 19 and a −3-dB bandwidth of 100 MHz. A 47-nH parasitic inductance due to the interconnections between the detector and the electronics has also been taken into account. Figure 9.7 reports a comparison between the output pulse that is produced by the detection system in response to a single photon and the corresponding simulation result that is obtained by using the described model with the extracted parameters: the fitting between measurement and simulation results is very good. In particular, the rising edge of the output pulse is well reproduced by the simulation.

TABLE 9.1

Parameters of the SiPM under Test

R_q	49.6 kΩ
V_{BR}	70.35 V
Q_{TOT}	160 fC
C_d	80.14 fF
C_q	15.49 fF
C_g	18.24 pF

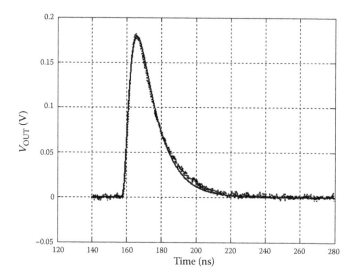

FIGURE 9.7   Fitting between measurement and simulations for the pulse produced by the SiPM (Hamamatsu S10931-050P).

9.3   FRONT-END ELECTRONICS FOR SiPMs

In typical applications, SiPMs can be employed to measure two quantities. The first one is the energy of the detected event, proportional to the total collected charge, i.e., to the total number of detected photons that trigger avalanche breakdown in a microcell of the SiPM. In this kind of measurement, as mentioned in Section 9.2, very often, the SiPM is coupled to a scintillating crystal, which converts an incident gamma photon, associated to the event to be detected, into a number of visible photons that are distributed in time according to the characteristic decay time of the scintillator. This measurement is affected by a number of errors that are caused by different phenomena that make the behavior of the detector nonideal. In particular, there is a finite probability that an incident photon is able to trigger an avalanche breakdown in one of the microcells of the SiPM, called photon detection efficiency; thus, not all the incident photons can be detected. So-called *dark pulses* are randomly generated at a given rate by microcells, which undergo an avalanche breakdown in the absence of incident photons. During an avalanche triggered in a microcell, photons can be emitted, which, in turn, can trigger a further avalanche phenomenon in a different microcell (optical crosstalk). The avalanche can be retriggered within the same microcell during the recovery phase, producing a total amount of charge that is higher than the one corresponding to the gain of the detector (afterpulsing). All the mentioned phenomena contribute to the inaccuracy of the energy measurement that is performed by means of an SiPM [5].

The second kind of measurement concerns the identification of the occurrence time of the event to be detected. Thanks to the operation based on avalanche breakdown, the current pulse generated by an SiPM exhibits a very short rise time, which makes the detector suitable for a vast range of applications requiring good timing accuracy. Single-photon detection capability is another interesting feature of the SiPM and can be exploited in frontier techniques, where time coincidence measurements between a trigger signal and the arrival of single detected photons must be performed with great accuracy, in the order of tens of picoseconds [8].

In both kinds of measurements, the architecture and performance of the front-end electronics used to read out the detectors must be carefully selected in order to achieve the desired specifications for a detection system that is based on SiPMs. In Sections 9.3.1 through 9.3.3, the advantages and drawbacks of the main solutions adopted for the front-end electronics to be coupled to SiPMs will be analyzed and discussed so as to give some guidelines for the most suitable choice depending on the requirements of the application. Our analysis will focus on integrated implementation of the proposed circuits, since, in the vast majority of the applications, detection systems with large numbers of channels are required, and the recourse to integrated solutions is mandatory to guarantee compactness and reliability.

9.3.1 CHARGE-SENSITIVE AMPLIFIER READOUT

First, let us consider the most widespread architecture that is used to read out particle detectors, i.e., the classic charge-sensitive amplifier (CSA). The well-known basic CSA configuration is depicted in Figure 9.8: a low-noise, inverting voltage amplifier, usually based on a cascode or folded cascode stage, is configured as an integrator by means of a feedback capacitance C_F, which collects the charge Q_{TOT} that is generated by the detector in response to an event and produces an output voltage that is proportional to Q_{TOT}:

$$V_{OUT} = \frac{Q_{TOT}}{C_F}. \tag{9.6}$$

The main drawback of this kind of readout approach when energy measurements have to be done is the limitation in terms of dynamic range. Since the charge released

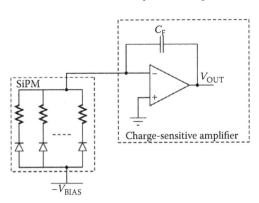

FIGURE 9.8   SiPM readout by means of a CSA.

by a single microcell of an SiPM undergoing an avalanche breakdown is in the order of hundreds of femtocoulombs, and the dynamic range of the V_{OUT} is limited by the power supply voltage V_{DD}, the absolute maximum number of photons that can be detected without saturating the preamplifier depends on the value of the capacitance C_F. Let us consider, for instance, a standard 130-µm CMOS technology, representing a good compromise between a possible level of integration and cost, with a typical V_{DD} voltage of 1.2 V. The possible dynamic range of the CSA output would be around 1 V, and, if the charge corresponding to 100 microcells should be integrated, a feedback capacitance of 10 pF would be needed, which is a huge value. This example does not correspond to a worst case, since, for instance, in PET application, a photoelectric interaction of a 511-keV photon within a lutetium yttrium orthosilicate (LYSO) scintillator produces approximately 1000 photons. Integration of such large capacitance values is impractical, due to the required amount of silicon area, especially when the number of front-end channels to be placed in a single ASIC are remarkable. Considering also the large equivalent capacitance of the detector, in the order of some tens or even hundreds of picofarads, depending on the number and size of the microcells contained in the detector, the output of the gain stage must be able to drive large capacitive loads without introducing severe limitations to the frequency response of the front-end electronics. This implies that an output buffer with sufficient drive capabilities is required, which results in a remarkable increase of the power consumption that is needed to avoid the formation of a slow pole at the output of the circuit in Figure 9.8.

In the case of energy measurements, the equivalent input noise of the CSA does not represent a serious issue, thanks to the large gain of the SiPM; thus, the recourse to a shaping filter to optimize the signal/noise ratio is not needed. Usually, a resistor R_F parallel to C_F is utilized to continuously reset the integration capacitance so as to guarantee the return to the baseline of the output signal and to establish a DC path for the leakage current of the detector. The time constant $\tau_F = R_F C_F$ must be slow enough to allow integration of the entire charge that is released by the detector in response to an event. Thus, in any case, τ_F should be adequately slower than the recovery time of the SiPM τ_r, and, in case a scintillator is used, its decay time constant τ_S must be taken into account, so τ_F should be sufficiently larger than the sum $\tau_S + \tau_r$.

In case time measurements are of interest, the typical solution for time pick-off is leading-edge discrimination. The output signal of the CSA is compared to a threshold V_{TH} by means of a fast leading-edge discriminator, which fires as soon as the signal overcomes the threshold, marking the arrival of a valid event. The parameter to be minimized to guarantee the best timing accuracy for the detection system is the time jitter σ_t, expressed by means of the well-known equation:

$$\sigma_t = \frac{\sigma_n}{\left.\dfrac{dV_{OUT}}{dt}\right|_{V_{OUT}=V_{TH}}}, \tag{9.7}$$

where σ_n is the root mean square (rms) output noise of the front-end electronics, and the denominator represents the slope of the output signal that is evaluated at the instant when the threshold is reached [15]. Of course, Equation 9.7 considers an ideal leading-edge discriminator, able to reproduce a sharp transition of its output signal exactly when the threshold is overcome. In other words, for timing accuracy, the noise/slope ratio should be minimized; thus, in this case, both the noise and frequency response performance of the front-end are relevant. Moreover, in the following, the response of the detector coupled to the front-end electronics to a single-photon event will be considered, since, if the timing accuracy of this response is optimized, the detection system will exhibit optimal timing performance also in case the event to be detected causes the production of more photons.

In Figure 9.9, the CSA has been coupled to the equivalent model of the SiPM that was introduced in Section 9.2.

As pointed out in Section 9.2, due to the effects of the capacitance C_q, a fraction Q_0 of the total charge Q_{TOT} is collected very quickly by the external circuit. If the avalanche current is modeled by means of a Dirac's delta, as in our model, Q_0 is collected istantaneously and is given by:

$$Q_0 = \Delta V C_q = Q_{TOT} \frac{C_q}{C_d + C_q}. \tag{9.8}$$

Equation 9.8 holds true since the voltage at node B in Figure 9.9 is very small as compared to the voltage at node A, thus the potential at node B can be considered a virtual ground.

In fact, the total impedance seen between node B and ground in the SiPM model is much smaller than the impedance that is given by R_q and C_q in parallel. Since, in the very first phase of the fast transient following the triggering of the avalanche breakdown, the only relevant elements in the circuit of Figure 9.10

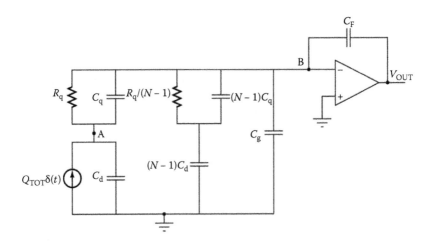

FIGURE 9.9 Electric model of the SiPM coupled to the CSA.

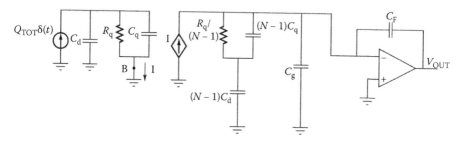

FIGURE 9.10 Equivalent, simplified version of the circuit in Figure 9.9.

are the capacitances; the charge Q_0 is immediately localized on the equivalent capacitance C_{eq}:

$$C_{eq} = C_g + (N-1)\frac{C_d C_q}{C_d + C_q} \cong C_g + N\frac{C_d C_q}{C_d + C_q}. \tag{9.9}$$

In Equation 9.9, the input capacitance of the gain stage has been neglected, since it is much smaller than C_{eq}. Note that the rest of the total charge released by the detector, i.e., $C_d \Delta V$, is collected according to the slow recovery time constant τ_r; thus, it cannot be conveniently exploited for fast-timing applications, since the slope of the associated component of the output pulse is very small. As a consequence, a good timing performance can be achieved only by exploiting the fast component Q_0 of the total charge Q_{TOT}. Therefore, from now on, we will focus on the response of the front end to this component. For this purpose, we can refer to the high-frequency equivalent circuit that is shown in Figure 9.11.

Given the gain–bandwidth (GBW) product of the inverting voltage amplifier shown in Figure 9.11, expressed in radians per second, the bandwidth of the CSA is:

$$\omega_F = \text{GBW}\frac{C_F}{C_{eq} + C_F}. \tag{9.10}$$

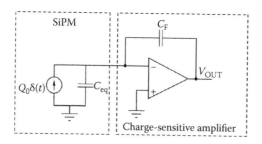

FIGURE 9.11 High-frequency approximation of the circuit in Figure 9.10.

Equation 9.10 shows that ω_F is strongly affected by the value of the equivalent capacitance C_{eq} of the detector: in case timing accuracy is of interest, SiPMs with a large number of microcells cannot be adopted, and, as a consequence, the number of readout channels must be increased at the cost of increased system complexity and power consumption. Exploiting Equation 9.10, the maximum slope of the output signal of the CSA, evaluated at $t = 0$, is

$$\left.\frac{dV_{OUT}}{dt}\right|_{MAX} = \frac{Q_0}{C_F}\omega_F = \frac{Q_0}{C_{eq} + C_F}\text{GBW}. \tag{9.11}$$

Concerning the contribution of the noise to the timing accuracy, if the spectral density of the equivalent input voltage noise is e_n^2, expressed in square voltages per hertz, an accurate evaluation of the rms value of the output voltage noise σ_n in analytical form is very complicated for the circuit in Figure 9.10. To have an idea of the absolute best performance that can be achieved by the circuit, the lower limit of σ_n can be estimated considering again the detector that is simply modeled by the capacitance C_{eq} that was defined in Equation 9.9, and as depicted in Figure 9.12.

Actually, this assumption describes with good accuracy the noise behavior of the system SiPM + CSA only at high frequency, since, at low frequency, a realistic estimation of the equivalent capacitance of the detector is the following:

$$C_{eq\_LF} = C_g + NC_d. \tag{9.12}$$

In fact, at low frequency, the impedance associated to the photodiode capacitance C_d prevails in the series with the parallel $R_q // C_q$, which can be considered a short circuit. Thus, at low frequency, the noise gain becomes $1 + C_{eq\_LF}/C_F$, larger than $1 + C_{eq}/C_F$. Since our purpose is only an approximated noise analysis that is carried out to find out the absolute best theoretic timing performance that can be achieved when the CSA approach is used, we will also consider C_{eq} as the equivalent capacitance of the detector for the estimation of σ_n, as illustrated in Figure 9.12. The rms output

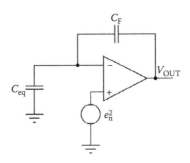

FIGURE 9.12   CSA readout: Estimation of the effects of electronic noise.

noise is given by the following equation, where the expression 9.10 has been used for the closed-loop bandwidth ω_F:

$$\sigma_n \cong \sqrt{\int_0^\infty e_n^2 \left(1 + \frac{C_{eq}}{C_F}\right)^2 \frac{1}{\left|1 + j\frac{\omega}{\omega_F}\right|^2} \, df} = \frac{e_n}{2} \sqrt{\left(1 + \frac{C_{eq}}{C_F}\right) GBW}. \tag{9.13}$$

Last, by applying Equation 9.7, the expression of the absolute minimum time jitter results in

$$\sigma_t \cong \frac{e_n(C_{eq} + C_F)}{2Q_0} \sqrt{\left(1 + \frac{C_{eq}}{C_F}\right) \frac{1}{GBW}}. \tag{9.14}$$

From Equation 9.14, we can conclude that, in case a CSA is used to read out the SiPM, the timing accuracy is strongly limited by the large equivalent capacitance of the detector C_{eq}. As pointed out in the conclusions of Section 9.3.1, SiPMs with a limited number of microcells must be used if we want to reduce the time jitter. Some advantages can be obtained by increasing the feedback capacitance, which would also be helpful for the dynamic range, in case energy measurements must be done. According to Equation 9.14, σ_t exhibits a minimum for $C_F = C_{eq}/2$, but this value of the integration capacitance is too large for typical SiPMs and thus impractical. Furthermore, the increase of C_F would cause a decrease of the gain of the CSA and, as a consequence, very small amplitudes of the output pulse that is produced by the single microcell. Therefore, the fast discriminator that must compare this signal with the threshold cannot be considered ideal at all, and the noise and jitter introduced by this circuit must be adequately taken into account.

The final remarks that can be drawn for the CSA approach are the following:

- When energy measurements must be carried out, the circuit is affected by severe dynamic range limitations, especially when modern deep-submicron technologies are employed.
- For timing measurements, only small SiPMs with a limited number of microcells can be employed, since for large equivalent detector capacitances, both the noise and slope of the output signal are penalized.
- If small-area SiPMs are used for timing, the feedback capacitance should be increased to minimize the output noise and improve the jitter, but this causes a decrease of the gain so that the comparator cannot be considered ideal, and its contributions to the jitter must be considered.

9.3.2 Voltage-Mode Readout

Very often, the SiPM is read out by exploiting the configuration that is shown in Figure 9.3 [16], which has also been used in Section 9.2 to illustrate the effectiveness of the electric model that is proposed for the detector. Figure 9.13 shows this model that is coupled to the resistor R_{IN}, which converts the current pulse of the detector into a voltage, considering the same approximation that was used in Figure 9.10.

The voltage across the resistor R_{IN}, i.e., the input voltage of the amplifier $V_{IN}(t)$, contains two contributions. The first contribution is due to the *fast* charge Q_0, which is almost immediately collected on the total equivalent capacitance C_{eq} at the input of the voltage amplifier, and the second contribution is associated to the rest of the charge that is released by the microcell, i.e., the *slow* component $C_d \Delta V$. Considering the first contribution, when an avalanche breakdown is triggered into a microcell of the detector, the input voltage of the amplifier $V_{IN}(t)$ exhibits a sudden variation to the value:

$$V_{IN}(0) = \frac{Q_0}{C_{eq}}. \tag{9.15}$$

Then, the charge Q_0 starts to flow through the resistor R_{IN}, and the voltage V_{IN} decreases with a transient behavior that is ruled by two time constants τ_1 and τ_2, as shown in Figure 9.14 for two different values of R_{IN}. The SiPM parameters used in the simulations are the same as those that were reported in Table 9.1 for the Hamamatsu S10931-050P.

An approximated analysis yields the following values for the time constants:

$$\tau_1 \cong \tau_r + R_{IN}(C_g + NC_d) \quad \tau_2 \cong \frac{1}{\dfrac{1}{R_{IN}C_{eq}} + \dfrac{1}{\tau_r}\dfrac{C_g + NC_d}{C_{eq}}}. \tag{9.16}$$

From Equation 9.16, it is apparent that the slow time constant τ_1 depends not only on the intrinsic recovery time of the SiPM τ_r but also on the resistor R_{IN}. If the parasitic capacitance C_g is in the same order of magnitude of NC_q, which is a reasonable assumption, the contribution $R_{IN}(C_g + NC_d)$ is equivalent to an increase of the quenching resistance that is equal to NR_{IN}. As a consequence, if the value of R_{IN} is increased, the tail of the signal V_{IN} becomes slower, and the effective recovery time of

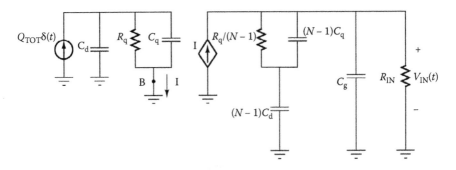

FIGURE 9.13   Electric model of the SiPM coupled to the resistor R_{IN}.

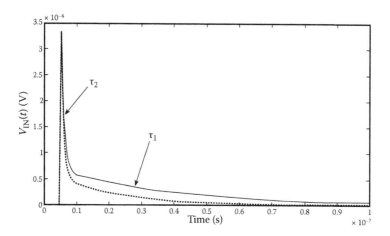

FIGURE 9.14 Contribution of Q_0 to the input voltage of the amplifier $V_{IN}(t)$ (dotted: R_{IN} = 50 Ω, solid: R_{IN} = 100 Ω).

the detection system grows, as shown in Figure 9.14. On the one hand, this can cause pileup and errors in energy measurements, which limit the maximum event rate that can be handled; on the other hand, the pileup also affects the accuracy of time mea-surements due to fluctuations of the instant in which the threshold is overcome by the signal with respect to the time of the event that is caused by variations of the baseline. Thus, the low values of R_{IN} are mandatory to avoid this kind of problems.

For low values of R_{IN}, the time constant τ_2, which characterizes the fast fall of the signal that is associated to the charge Q_0, can be further approximated as

$$\tau_2 \cong R_{IN}C_{eq}. \tag{9.17}$$

Considering the second contribution to $V_{IN}(t)$, given by the fraction $Q_d = C_d\Delta V$ of the total charge Q_{TOT}, the related component in the current $I(t)$ in Figure 9.13 is $I_d(t)$:

$$I_d(t) = \frac{Q_d}{\tau_r}e^{-\frac{t}{\tau_r}} = \frac{\Delta V}{R_q}\frac{C_d}{C_d+C_q}e^{-\frac{t}{\tau_r}}. \tag{9.18}$$

The resulting contribution to $V_{IN}(t)$ exhibits a rising edge that is ruled by the time constant τ_2 and a tail that is dominated in practice by the time constant τ_1, slower than τ_r, as shown in Figure 9.15. As a consequence, the relevant component of the signal $V_{IN}(t)$ that can be exploited for accurate timing application is once again asso-ciated to the fast charge Q_0.

As already discussed in this section, the low values of the resistor R_{IN} are conve-nient in increasing the maximum event rate, which can be accepted without affecting the accuracy of both time and energy measurements. If we focus on energy measure-ments, the bandwidth of the voltage amplifier used is not of much importance, since, in case the photons arrive all at the same time on the detector, the peak of the voltage

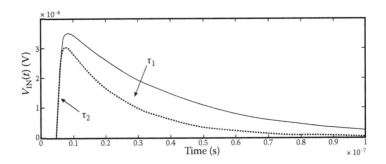

FIGURE 9.15 Contribution of $I_d(t)$ to the input voltage of the amplifier $V_{IN}(t)$ (dotted: R_{IN} = 50 Ω, solid: R_{IN} = 100 Ω).

signal $V_{OUT}(t)$ obtained at the amplifier output is proportional to the total charge that is released by the detector, since the system is linear. In case a scintillator is used, V_{OUT} must be integrated by means of a low-pass filter with a suitable time constant, chosen according to the sum of the time constant τ_1 and the decay time constant of the scintillator, in order to provide a signal that is proportional to the total charge that is associated to the scintillation event.

Let us now consider the evaluation of the time jitter that is associated to the voltage-mode readout approach for the SiPM. As already explained, we can only focus on the contribution due to the charge Q_0 and, in particular, on the very first part of the transient behavior of $V_{IN}(t)$, for $t \ll \tau_1$, considering in practice the response of the amplifier with gain A_V and bandwidth BW = $1/\tau_A$ to the following approximated expression of $V_{IN}(t)$:

$$V_{IN}(t) \cong \frac{Q_0}{C_{eq}} e^{-\frac{t}{\tau_2}}. \tag{9.19}$$

The resulting output response of the amplifier is the following:

$$V_{OUT}(t) \cong A_V \frac{Q_0}{C_{eq}} \frac{\tau_2}{\tau_2 - \tau_A} \left(e^{-\frac{t}{\tau_2}} - e^{-\frac{t}{\tau_A}} \right). \tag{9.20}$$

Also, in this case, the absolute maximum slope of the output signal corresponds to $t = 0$:

$$\left. \frac{dV_{OUT}(t)}{dt} \right|_{MAX} \cong A_V \frac{Q_0}{C_{eq}} \frac{1}{\tau_A} = GBW \frac{Q_0}{C_{eq}}, \tag{9.21}$$

where GBW represents, as usual, the gain–bandwidth product of the voltage amplifier A_VBW, where BW is always expressed in radians per second. From Equation 9.21, we can conclude that on the one hand, to a first approximation, the maximum slope of $V_{OUT}(t)$ does not depend on the value of R_{IN}. On the other hand, the peak value of $V_{OUT}(t)$ is reported in the following expression:

$$V_{PEAK} \cong A_V \frac{Q_0}{C_{eq}} \frac{\tau_2}{\tau_2 - \tau_A} \left(\left(\frac{\tau_2}{\tau_A}\right)^{\frac{\tau_A}{\tau_A - \tau_2}} - \left(\frac{\tau_2}{\tau_A}\right)^{\frac{\tau_2}{\tau_A - \tau_2}} \right). \quad (9.22)$$

V_{PEAK} is an increasing function of τ_2 and R_{IN} (see Figure 9.16), and, as a consequence, the amplitude of the signal at the input of the voltage comparator, which generates the trigger signal, becomes larger if R_{IN} is increased. Moreover, the slope of $V_{OUT}(t)$ around the threshold, which is smaller than the maximum slope that is expressed by Equation 9.21 and evaluated at $t = 0$, slightly increases for increasing values of R_{IN}. Thus, for timing applications, R_{IN} should be increased to guarantee optimal operating conditions for the comparator and, at the same time, to slightly increase the slope of the output signal around the threshold. However, the above-mentioned limitations in terms of maximum event rate when the resistance R_{IN} is increased must always be taken into account.

The evaluation of the rms value of the output noise for the voltage-mode approach is straightforward:

$$\sigma_n = \frac{e_n A_V}{2} \sqrt{BW}. \quad (9.23)$$

As a consequence, the time jitter has the following inferior limit:

$$\sigma_t = \frac{e_n A_V \sqrt{BW}}{2GBW \dfrac{Q_0}{C_{eq}}} = \frac{e_n C_{eq}}{2Q_0} \sqrt{\frac{1}{BW}}, \quad (9.24)$$

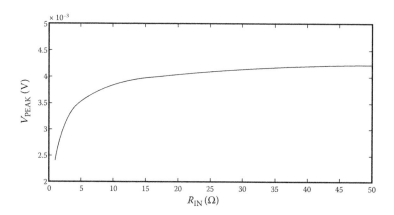

FIGURE 9.16 Peak value of the fast component of $V_{OUT}(t)$ (contribution of Q_0) as a function of R_{IN}.

which represents an improvement with respect to the corresponding value that is associated to the CSA approach, reported in Equation 9.14.

In conclusion, for timing applications, the voltage-mode readout exhibits better performance with respect to the CSA solution, and the main drawback is represented by limitations in the maximum event rate, which can be managed when the resistance R_{IN} is increased to approach the best conditions for what concerns timing accuracy. Of course, improvements of the timing performance are obtained by increasing the bandwidth of the amplifier.

9.3.3 CURRENT-MODE READOUT

In the voltage-mode approach discussed in Section 9.3.2, the main goal was reproducing as best as possible at the output of the circuit, the waveform of the current pulse that is generated by the detector, thus preserving its favorable features, without introducing much loading effects, mainly due to the resistor R_{IN}, and using a wide-bandwidth amplifier. To approach the ideal operating conditions for the detector, obtained when the bias voltage is directly connected at its terminal, as shown in Figure 9.3, the resistance R_{IN} should be as low as possible, but, as discussed in Section 9.3.2, this is not possible when the voltage-mode approach is adopted. As a consequence, if the SiPM is read out by means of a current buffer with very low input resistance R_{IN}, the detector will work in operating conditions that are very close to the ideal case. The output current of the buffer can be easily replicated using simple current mirroring techniques and suitably scaled in order to fulfill the requirements in terms of dynamic range. One of the replicas can be used to form the trigger signal, either directly using a fast leading-edge current discriminator or converting the current into a voltage and using a voltage comparator. Another output current of the buffer, suitably scaled down, can be sent to an integrator, which provides the energy information, if required. Figure 9.17 illustrates a typical arrangement, showing a two-output current buffer, with the *fast path*, used for time measurements, and the *slow path*, which provides an output that is proportional to the total integrated charge, and thus to the energy that is associated to the detected event [17].

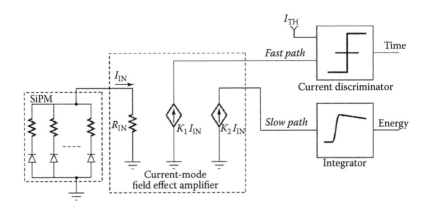

FIGURE 9.17   Example of current-mode readout of the SiPM.

Let us now analyze the timing performance that can be achieved by exploiting the current-mode approach. We are now interested to an approximated expression of the current $I_{IN}(t)$, which flows into the input resistance R_{IN} of the current buffer, which is very low. By exploiting the analysis carried out for the voltage-mode approach, considering only the fast fraction of the charge that is released by the detector during the avalanche breakdown and the very first part of the transient behavior, we obtain

$$I_{IN}(t) \cong \frac{Q_0}{R_{IN}C_{eq}} e^{-\frac{t}{\tau_2}} \cong \frac{Q_0}{\tau_2} e^{-\frac{t}{\tau_2}}. \tag{9.25}$$

If the bandwidth of the fast path is $BW = 1/\tau_A$ and we directly exploit a current discriminator to compare the output signal of the current buffer with a threshold, the following expression for the output current signal is achieved:

$$I_{OUT}(t) \cong \frac{Q_0}{\tau_2 - \tau_A} \left(e^{-\frac{t}{\tau_2}} - e^{-\frac{t}{\tau_A}} \right). \tag{9.26}$$

In Equation 9.26, a unity gain has been assumed for the current buffer. The peak value of $I_{OUT}(t)$ can be easily evaluated as follows:

$$I_{PEAK} \cong \frac{Q_0}{\tau_2 - \tau_A} \left(\left(\frac{\tau_2}{\tau_A} \right)^{\frac{\tau_A}{\tau_A - \tau_2}} - \left(\frac{\tau_2}{\tau_A} \right)^{\frac{\tau_2}{\tau_A - \tau_2}} \right). \tag{9.27}$$

Figure 9.18 reports the behavior of I_{PEAK} as a function of R_{IN}.

It is apparent that in decreasing R_{IN}, i.e., decreasing the time constant τ_2, higher values of the output current peak are obtained. Moreover, the maximum slope of $I_{OUT}(t)$, always corresponding to $t = 0$, is

$$\left. \frac{dI_{OUT}(t)}{dt} \right|_{MAX} \cong \frac{Q_0}{\tau_2} \frac{1}{\tau_A} = \frac{Q_0}{\tau_2} BW. \tag{9.28}$$

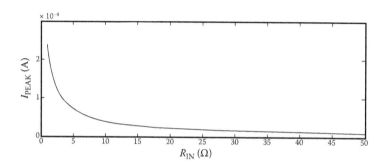

FIGURE 9.18 Peak value of the fast component of $I_{OUT}(t)$ (contribution of Q_0) as a function of R_{IN}.

Also, the maximum slope of the output signal increases for decreasing values of τ_2 and R_{IN}; thus, we can conclude that the low values of R_{IN} contribute to improve the timing performance of the current-mode readout approach and decrease the time constant τ_1, which characterizes the tail of the output pulse. This represents an advantage of the current-mode approach over the voltage-mode approach.

Concerning noise, we refer to the simplest CMOS implementation of a current buffer in the form of a common gate amplifier, represented in Figure 9.19.

The main contributions to the noise current at the output of the circuit are due to the common gate metal–oxide-semiconductor field-effect transistor (MOSFET) M_1 and to the current sources I_{BIAS} and I_{TH}, realized by means of metal–oxide-semiconductor transistors. In an approximated analysis, the main noise contribution comes from the input transistor, since the size of the MOSFETs used to realize I_{BIAS} and I_{TH} can be chosen so as to make their transconductance much smaller than the transconductance g_{m1} of the input transistor M_1. In fact, in the circuit represented in Figure 9.19, the resistance R_{IN} is given by $1/g_{m1}$; thus, g_{m1} must be large in order to minimize R_{IN} and, as a consequence, the noise current spectral density $i_{n1}^2 = 4kT\gamma g_{m1}$ (expressed in square amperes per hertz) associated to M_1 is dominant. Actually, this noise current can reach the current discriminator only at high frequencies, when the equivalent capacitance of the detector bypasses to ground the source of M_1, but if we neglect this consideration and integrate i_{n1}^2 over all the spectrum of frequencies, we will obtain a conservative, worst-case expression of the rms value of the noise current σ_n at the input of the discriminator:

$$\sigma_n \cong \sqrt{\int_0^\infty i_{n1}^2 \frac{1}{\left|1 + j\dfrac{\omega}{BW}\right|^2} \, df} = \frac{i_{n1}}{2}\sqrt{BW}. \tag{9.29}$$

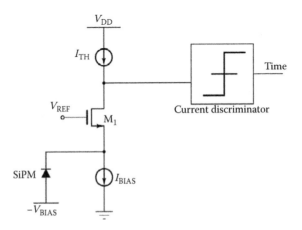

FIGURE 9.19 Common gate M_1 used as current buffer.

Eventually, a simple expression of the time jitter can be written exploiting Equations 9.28 and 9.29:

$$\sigma_t \cong \frac{i_{n1}\sqrt{BW}}{2\dfrac{Q_0}{\tau_2}BW} = \frac{i_{n1}R_{IN}C_{eq}}{2Q_0}\sqrt{\frac{1}{BW}} = \frac{i_{n1}C_{eq}}{2g_{m1}Q_0}\sqrt{\frac{1}{BW}} = \frac{e_{n1}C_{eq}}{2Q_0}\sqrt{\frac{1}{BW}}, \quad (9.30)$$

where $e_{n1}^2 = i_{n1}^2/g_{m1}^2 = 4kT\gamma/g_{m1}$ is the gate-referred equivalent noise voltage of the input MOSFET M_1. The expression 9.30 of the time jitter for the current buffer in Figure 9.19 is formally similar to Equation 9.24, obtained for the voltage-mode approach. The only difference is that in Equation 9.30, the gate-referred noise voltage e_{n1} of M_1 replaces the equivalent input noise e_n of the voltage amplifier. Moreover, large-bandwidth current buffers are much simpler and easier to be designed than voltage amplifiers with the same speed performance.

Anyway, in case the current buffer approach is employed, on the basis of the previous discussion, we can conclude that the best operating conditions for both the detector and the comparator can be easily approached by decreasing the input resistance of the buffer R_{IN}, avoiding at the same time an undesired increase of the time constant, which characterizes the tail of the current pulse that is produced by the detector. As a consequence, the results of our analysis demonstrate the advantages of the current-mode solution over the other readout techniques for SiPM detectors. This explains the success of this approach and the large number of application examples of the current-mode technique in the literature. For instance, an open-loop implementation of the input current buffer can be found in [18], whereas many examples of circuit configurations that exploit negative feedback to decrease the input resistance of the buffer exist [19–21]. On the other hand, realizations based on the voltage-mode approach can be found, for instance [22].

9.4 ARCHITECTURE EXAMPLES

In this section, two examples of ASIC architectures designed for the readout of an array of SiPMs will be presented. The different solutions used for the front end, always based on the current-mode approach, the techniques used for energy and time measurements, and the main features of the overall organization of the ASICs will be discussed.

The first example is the family of CMOS ASICs called BASIC [14,23,24]. In this case, the front end is a current buffer that is based on a feedback structure, represented in Figure 9.20.

The common gate transistor M_1 is enclosed in a feedback loop, which encompasses the common source M_2 and the diode-connected MOSFET M_4, to increase the bandwidth and to decrease the input impedance of the circuit. Since the two dominant open-loop poles of the circuit are not far apart, due to the large equivalent detector capacitance, the loop gain has been limited to approximately 10 to prevent stability problems. M_3 is used to tune the loop gain by setting appropriate bias currents in the MOSFETs. An interesting feature of the circuit, common with

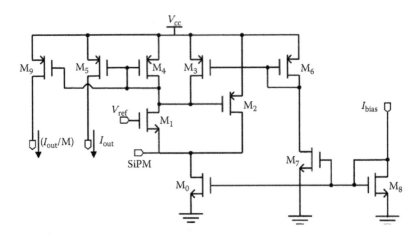

FIGURE 9.20   Input current buffer of BASIC.

many different realizations of the current buffer, is the fact that if the detector is DC coupled to the electronics, it is possible to fine-tune its bias voltage, and thus its gain, by varying the voltage V_{ref} at the gate of the input transistor. For V_{ref} in the range between 1 and 2 V, the current buffer exhibits a 250-MHz bandwidth and an input impedance of approximately 17 Ω, with a current consumption of 800 µA at the standard 3.3-V power supply voltage of the CMOS 0.35-µm technology. The buffer has two output currents, which are differently scaled replicas of the SiPM current pulse: as pictured in Figure 9.21, the first replica, not scaled down, is sent directly to the input of a current discriminator to generate the signal that marks the arrival of a valid event, i.e., the trigger signal. The other replica of the detector current, suitably scaled down by a factor M to cope with the above-mentioned dynamic range issues, is integrated by means of a CSA with variable gain and dumping time constant. A peak detector holds the peak voltage at the CSA output and works as an analog memory during the readout procedures. Finally, a baseline holder circuit encloses the CSA in a very slow feedback loop, useful to set the DC value of the CSA output at V_{BL} = 300 mV, without affecting the signal waveform.

The current discriminator is based on the principle that is illustrated in Figure 9.22: the threshold current I_{th} flows into the NMOS M_1, thus forcing the output of the inverting amplifier V_{trig} at a high level. As soon as the current I_{OUT} overcomes the threshold, M_1 is switched off, M_2 is turned on, and the output of the amplifier suddenly goes low, thus marking the arrival of a valid event. A fall time of the output signal of approximately 300 ps has been achieved.

The described analog channel has been designed in a standard 0.35-µm CMOS technology, and the target application is the readout of SiPM matrices that are coupled to pixelated scintillators in PET detectors. In the last version of the circuit, the dynamic range of the input charge is approximately 130 pC. The energy resolution of the ^{22}Na spectrum obtained using the analog channel of BASIC to read out a 3 × 3-mm^2 SiPM coupled to a 3 × 3 × 10-mm^3 LYSO crystal is approximately 11%.

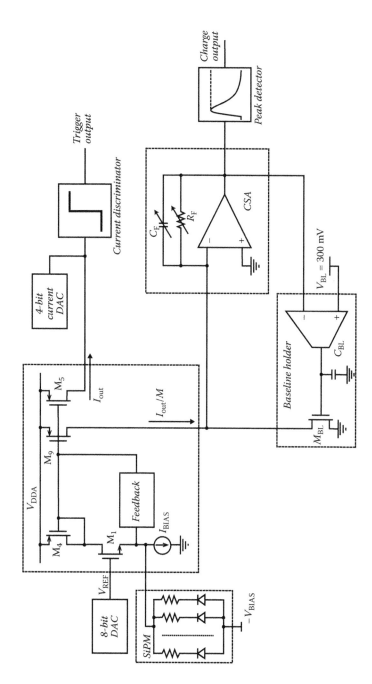

FIGURE 9.21 Architecture of the analog channel of BASIC.

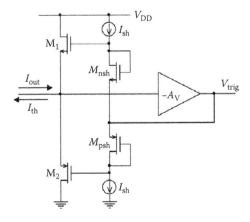

FIGURE 9.22 Working principle of the current discriminator.

Concerning the architecture of the ASIC, a 32-channel version has been designed according to the architecture that is illustrated in Figure 9.23.

The analog outputs of the 32 channels are multiplexed toward an external pad: the analog multiplexer (MUX) is controlled by the internal logic block, which automatically starts the multiplexing of the channels synchronously with the clock in sparse or in serial mode, according to the chosen configuration, triggered by the detection of a valid event. The trigger signal is produced by combining in fast-OR the outputs of the current discriminators of each channel. The time jitter measured considering the same source, detector, and scintillator that was mentioned in Section 9.3 for the measurements of energy resolution has been 1.2-ns full width half maximum (FWHM), which is sufficient for PET but not suitable for ToF-PET.

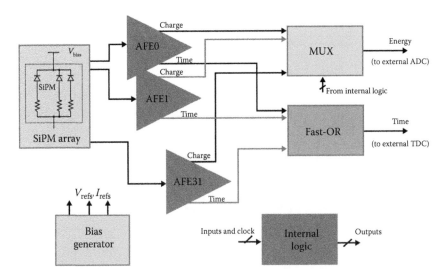

FIGURE 9.23 Architecture of the 32-channel ASIC.

Another version of the ASIC is based on a slightly different organization of the analog channel and the overall architecture, illustrated in Figure 9.24.

Here, the threshold that discriminates a valid event is not set on the amplitude of the current generated by the detector, but rather on the amount of charge that is contained in the detector pulse. Therefore, the current discriminator placed on the replica of the detector current provided by the input buffer (see Figure 9.21) is replaced by a voltage discriminator, which compares the output voltage of the charge-sensitive preamplifier to a given threshold. This arrangement of the analog channel is suitable when a nonpixelated scintillator is coupled to an array of SiPMs so that very often, the amount of photons collected by a single detector can be very low, and, as a consequence, the associated current pulse can have an amplitude that is comparable to one of a single dark pulse. If the architecture in Figure 9.21 is used, since the threshold of the current discriminator is set to get rid of the dark pulses, when this happens, the related channels have an output signal below the threshold, thus they are ignored in sparse readout even though they are associated to a non-negligible amount of charge.

The trigger signal obtained with the voltage discriminator's architecture, depicted in Figure 9.24, is intrinsically slower than the one that is provided by the current discriminator, which fully exploits the current-mode approach. In order to preserve the advantages offered by this approach, the available replica of the output current of the input buffer, scaled down by a factor N, is sent to a fast-summing circuit, which produces an output current that is proportional to the sum of the currents that are generated by the 32 detectors that are read out by the ASIC. It is very likely that the amplitude of the resulting signal, associated to the total number of incident photons on the surface covered by the 32 SiPMs, will be sufficiently large enough to overcome a threshold set for avoiding triggers from multiple dark pulses. Thus, a very fast trigger signal is obtained by a current discriminator that compares the output of the summing circuit to a suitable threshold current I_{TH}, as schematically illustrated in Figure 9.24. In the sparse readout operations, only the channels whose voltage comparator has fired are considered. An internal, 8-bit two-step flash analog-to-digital converter has also been added, and a further slow-trigger signal is also available, generated by the OR of the voltage comparators of the channels.

The timing accuracy achieved with the analog channel of BASIC chips is not sufficient for applications such as ToF-PET, since, in order to fully exploit the advantages offered by the ToF technique, time must be measured with a resolution of hundreds of picoseconds. The second example of ASIC architecture we propose has been designed to fulfill the requirements of such applications, considering once again systems where the detector is based on a continuous slab of scintillating crystal that is coupled to an array of SiPMs.

To satisfy these requirements, the detection threshold must be decreased down to the level of the signal that is associated to the single photoelectron, i.e., the single-fired microcell, to reduce as much as possible the contribution to the signal fluctuations due to the statistic distribution of the arrival times of the photons [7]. As a consequence, on the one hand, the noise level of the front-end electronics must be low enough to allow the required decrease of the threshold level. On the other hand, an effective solution must be put into practice to get rid of the triggers coming from the dark pulses that are generated by the SiPM.

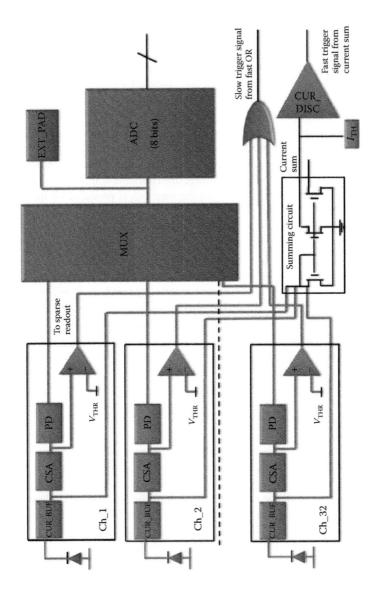

FIGURE 9.24 New architecture of the 32-channel ASIC.

TOT_AL is a four-channel prototype ASIC, designed in a standard 0.35-μm Silicon-Germanium (SiGe) technology, which tries to address these issues [25]. The architecture of the single channel, shown in Figure 9.25, consists of three main blocks: (1) the fast path, (2) the slow path, and (3) the control unit.

The fast path generates an accurate trigger for ToF measurements with a jitter of 235-ps FWHM. The slow path validates the input signal, rejecting the SiPM dark noise, and, for the acknowledged pulses, carries out the charge measurement. The control unit encloses a 43-bit configuration shift register, used to tune the programmable parameters of the analog channel and to manage the threshold adjustment. It also generates all the internal control signals and the output signal of the channel, which is a digital low voltage differential signaling (LVDS) pulse whose rising edge marks the arrival of a valid event with the required timing accuracy and whose duration is proportional to the associated energy.

The very first stage, a low-noise transimpedance amplifier (Z Amplifier in Figure 9.25), converts the SiPM current pulse into a voltage. In order to fulfill the timing requirements, the chosen amplifier structure is based on an open-loop current buffer, schematically represented in Figure 9.26.

The main component is the SiGe Heterojunction Bipolar Transistor (HBT) Q_1 that is configured as a common base, and the SiPM current is converted into a voltage by a polysilicon resistor. A wide swing cascode current mirror biases the stage, and an OP AMP, closed in a unity feedback, drives the HBT base voltage. Both the bias current and the base voltage can be adjusted: current tuning can compensate the load resistor variation due to process fluctuations that induce a bias change in the entire fast path. The HBT base voltage control allows a fine regulation of the emitter voltage and thus of the SiPM bias (Figure 9.26).

This architecture takes advantage from the SiGe technology that, instead of a standard CMOS, allows exploiting the higher-frequency performances of the HBT. Thanks to this device, low values of the input resistance can be achieved with a lower bias current than a MOSFET, thus reducing both the power consumption and the noise of the stage. With 1 mA of bias current, the input resistance obtained is approximately 25 Ω, and the signal/noise ratio is 25 dB for the signal corresponding to the single SiPM microcell fired (Hamamatsu S10931-050P).

The output voltage of the transimpedance amplifier is applied to a gain stage in the fast path and to a transconductance amplifier in the slow path. The first one is a large-bandwidth voltage amplifier that drives the input of a fast-voltage comparator, which compares the amplified signal with an adjustable threshold and produces a leading edge that accurately measures the arrival time of the γ-photon with a low jitter. The transconductance amplifier of the slow path (Y Amplifier in Figure 9.25) reconverts the first-stage output voltage into two current signals, which are scaled copies of the SiPM signal. These two current replicas are read by the dark pulse rejection and the time-over-threshold blocks, respectively.

The channel operation is described in the following. When the comparator of the fast path fires, the control unit sets the LVDS output to the high level and brings the channel out of the trigger-awaiting state. The dark pulse rejection block starts integration of the input current signal, which lasts a short, adjustable

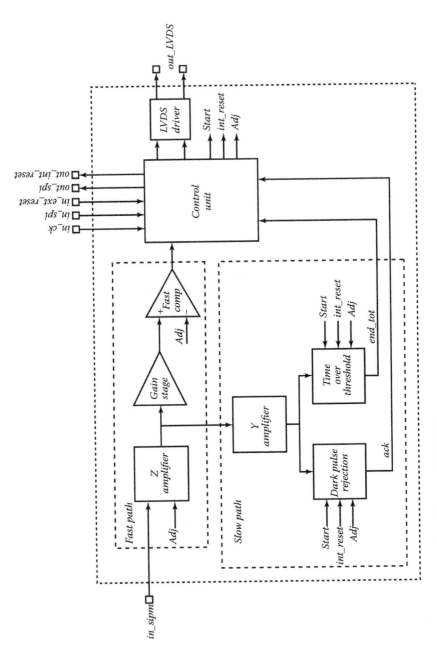

FIGURE 9.25   Analog channel architecture of TOT_AL.

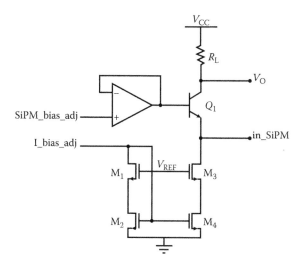

FIGURE 9.26 TOT_AL front-end architecture.

time window (TWA). At the end of TWA, if the integrated charge is lower than a programmable threshold, the signal is considered invalid, and the control unit produces an internal reset that sets back to zero the LVDS output and restores the channel's initial condition. In case the integrated charge overcomes the threshold at the expiration of the TWA, the signal is acknowledged as a valid event, and the readout operation goes on.

At the same time, when the fast comparator fires, the time-over-threshold block also starts the integration of the second replica of the current onto a capacitor. If the signal is recognized as valid, according to the mechanism just described, the integration goes on during a longer, adjustable time window (TWB). At the expiration of TWB, the integration is stopped, and the integration capacitance is discharged at a constant current. As soon as the voltage across the capacitance falls under a fixed threshold, the control unit generates a reset signal, and the LVDS output goes back to zero. In this case, the duration of the discharge phase is proportional to the total charge that is integrated during TWB.

In such a way, the leading edge of the LVDS output pulse always marks the arrival time of an event. In the case of an invalid event, the duration of the pulse is just the short TWA. On the other hand, when the event is recognized as valid by the dark pulse rejection block, the output pulse width is equal to the sum of TWB and the discharge time of the integration capacitance and hence is linearly related to the energy of the event. Thus, an accurate external TDC can be used to put a time stamp on both the leading and falling edge of the output pulse, so as (a) the occurrence time of the event is measured with the needed accuracy, (b) the events corresponding to a short duration of the output pulse, i.e., TWA, can be easily rejected, and (c) the energy associated to the valid event can be evaluated by means of the duration of the output pulse.

REFERENCES

1. G. Bondarenko, P. Buzhan, B. Dolgoshein, V. Golovin, E. Guschin, A. Ilyin, V. Kaplin et al. 2000. Limited Geiger-mode microcell silicon photodiode: New results. *Nuclear Instruments and Methods in Physics Research*, vol. A 442, no. 1–3, pp. 187–192.
2. P. Buzhan, B. Dolgoshein, L. Filatov, A. Ilyin, V. Kantzerov, V. Kaplin, A. Karakash et al. 2003. Silicon photomultiplier and its possible applications. *Nuclear Instruments and Methods in Physics Research*, vol. A 504, no. 1–3, pp. 48–52.
3. V. Golovin, V. Saveliev. 2004. Novel type of avalanche photodetector with Geiger mode operation. *Nuclear Instruments and Methods in Physics Research*, vol. A 518, no. 1–2, pp. 560–564.
4. V. D. Kovaltchouk, G. J. Lolos, Z. Papandreou, K. Wolbaum. 2005. Comparison of a silicon photomultiplier to a traditional vacuum photomultiplier. *Nuclear Instruments and Methods in Physics Research*, vol. A 538, no. 1–3, pp. 408–415.
5. D. Renker. 2006. Geiger-mode avalanche photodiodes, history, properties and problems. *Nuclear Instruments and Methods in Physics Research*, vol. A 567, no. 1, pp. 48–56.
6. Technical Information—MPPC, MPPC modules. Hamamatsu Photonics K.K., Japan. Available at https://www.hamamatsu.com/resources/pdf/ssd/mppc_techinfo_e.pdf.
7. V. Ch. Spanoudaky, C. S. Levin. 2010. Photo-detectors for time of flight positron emission tomography (ToF-PET). *Sensors*, ISSN 1424-8220, vol. 10, no. 11, pp. 10484–10505.
8. C. J. Chunnilall, I. P. Degiovanni, S. Kück, I. Müller, A. G. Sinclair. 2014. Metrology of single-photon sources and detectors: A review. *Optical Engineering*, vol. 53, no. 8, pp. 1–17.
9. S. Cova, M. Ghioni, A. Lacaita, C. Samori, F. Zappa. 1996. Avalanche photodiodes and quenching circuits for single-photon detection. *Applied Optics*, vol. 35, no. 12, pp. 1956–1976.
10. S. Seifert, H. T. van Dam, J. Huizenga, R. Vinke, P. Dendooven, H. Löhner, D. R. Schaart. 2009. Simulation of silicon photomultiplier signals. *IEEE Transactions on Nuclear Science*, vol. 56, no. 6, pp. 3726–3733.
11. D. Marano, M. Belluso, G. Bonanno, S. Billotta, A. Grillo, S. Garozzo, G. Romeo et al. 2014. Silicon photomultipliers electrical model extensive analytical analysis. *IEEE Transactions on Nuclear Science*, vol. 61, no. 1, pp. 23–34.
12. N. Pavlov, G. Mæhlum, D. Meier. 2005. Gamma spectroscopy using a silicon photo-multiplier and a scintillator. *2005 IEEE NSS Conference Record*, pp. 173–180, Puerto Rico, October 23–29.
13. F. Corsi, A. Dragone, C. Marzocca, A. Del Guerra, P. Delizia, N. Dinu, C. Piemonte, M. Boscardin, G. F. Dalla Betta. 2007. Modelling a silicon photo multiplier (SiPM) as a signal source for optimum front-end design. *Nuclear Instruments and Methods in Physics Research*, vol. A 572, no. 1, pp. 416–418.
14. F. Corsi, M. Foresta, C. Marzocca, G. Matarrese, A. Del Guerra. 2009. ASIC development for SiPM readout. *Journal of Instrumentation*, vol. 4, article no. P03004, pp. 1–10. doi:10.1088/1748-0221/4/03/P03004.
15. H. Spieler. 2005. *Semiconductor Detector Systems*. New York: Oxford University Press.
16. J. Y. Yeom, R. Vinke, C. S. Levin. 2013. Optimizing timing performance of silicon photomultiplier-based scintillation detectors. *Physics in Medicine and Biology*, vol. 58, no. 4, pp. 1207–1220.
17. F. Corsi, M. Foresta, C. Marzocca, G. Matarrese, A. Del Guerra. 2008. Experimental results from an analog front-end channel for silicon photomultiplier detectors. *2008 IEEE NSS Conference Record*, pp. 2010–2014, Dresden, October 19–25.
18. F. Powolny, E. Auffray, S. E. Brunner, E. Garutti, M. Goettlich, H. Hillemanns, P. Jarron et al. 2011. Time-base readout of a silicon photomultiplier (SiPM) for time of flight positron emission tomography (TOF-PET). *IEEE Transactions on Nuclear Science*, vol. 58, no. 3, pp. 597–604.

19. A. Comerma, D. Gascón, L. Garrido, C. Delgado, J. Marín, J. M. Pérez, G. Martínez, L. Freixas. 2013. Front end ASIC design for SiPM readout. *Journal of Instrumentation*, vol. 8, article no. C01048, pp. 1–7. doi:10.1088/1748-0221/8/01/C01048..

20. M. D. Rolo, R. Bugalho, F. Gonçalves, G. Mazza, A. Rivetti, J. C. Silva, R. Silva, J. Varela. 2013. TOFPET ASIC for PET applications. *Journal of Instrumentation*, vol. 8, article no. C02050, pp. 1–9. doi:10.1088/1748-0221/8/02/C02050..

21. I. Sacco, P. Fischer, M. Ritzert, I. Peric. 2013. A low power front-end architecture for SiPM readout with integrated ADC and multiplexed readout. *Journal of Instrumentation*, vol. 8, article no. C01023, pp. 1–5. doi:10.1088/1748-0221/8/01/C01023.

22. S. Callier, C. de La Taille, G. Martin-Chassard, L. Raux. 2012. EASIROC, an easy and versatile readout device for SiPM. *Physics Procedia*, vol. 37, pp. 1569–1576.

23. A. Argentieri, F. Corsi, M. Foresta, C. Marzocca, A. Del Guerra. 2011. Design and characterization of CMOS multichannel front-end electronics for silicon photomultipliers. *Nuclear Instruments and Methods in Physics Research*, vol. A 652, no. 1, pp. 516–519.

24. F. Ciciriello, F. Corsi, F. Licciulli, C. Marzocca, G. Matarrese, E. Chesi, E. Nappi, A. Rudge, J. Seguinot, A. Del Guerra. 2013. BASIC32_ADC, a front-end ASIC for SiPM detectors. *2013 IEEE NSS Conference Record*, pp. 1–6, Seoul, South Korea, October 27–November 2, 2013.

25. F. Licciulli, F. Ciciriello, F. Corsi, C. Marzocca, M. G. Bisogni. 2013. TOT_AL: An ASIC for TOF and DOI measurement. *2013 IEEE NSS Conference Record*, pp. 1–6, Seoul, South Korea, October 27–November 2, 2013.

10 CMOS Image Sensors for Radiation Detection

Nicola Guerrini

CONTENTS

10.1 Introduction 237
    10.1.1 Monolithic CMOS 238
10.2 Radiation in Silicon 239
    10.2.1 Charge Collection 240
10.3 Radiation Hardness...243
    10.3.1 Effects of Radiation on CMOS..243
        10.3.1.1 Ionization Effects...243
        10.3.1.2 Displacement Effects 244
    10.3.2 Radiation Hardening by Design 245
10.4 Noise Sources 246
    10.4.1 Shot Noise...247
    10.4.2 Thermal Noise 247
    10.4.3 $1/f$ Noise...249
    10.4.4 Random Telegraph Signal Noise 249
10.5 CMOS Sensor Architecture..249
    10.5.1 Pixels...250
        10.5.1.1 3T..250
        10.5.1.2 4T..251
        10.5.1.3 In-Pixel Processing 252
    10.5.2 Row Addressing..253
    10.5.3 Sensor Readout 255
10.6 Trends and Applications 256
10.7 Conclusions...258
References...259

10.1 INTRODUCTION

We do not need an expert in the semiconductor industry to be able to acknowledge the impact of complementary metal–oxide semiconductor (CMOS) technology in the world around us in recent years. And, this success did not arise just from inventions but mainly from their application to different fields [1].

Early integrated electronics in the 1950s and 1960s was dominated by bipolar devices, and we had to wait for the 1980s for CMOS to be considered a technology for the future [1].

The dimension scaling predicted by Moore's Law, the low leakage current, low voltage, low power, low cost, and versatility are all of the factors that contributed to the success of CMOS technology.

Since their introduction in the 1970s [2], charge-coupled devices (CCDs) have been the technology of choice for digital cameras, but the growth of CMOS meant that it was only a matter of time before the CCDs would have to face serious competition.

In the early 1990s, a research group in Jet Propulsion Laboratory (JPL) led by Eric Fossum developed the first CMOS image sensor [3]. The sensor was indeed fabricated using a standard CMOS process, but this is not all of the story. Using Eric Fossum's words, "*In essence, a small surface-channel CCD has been fabricated in each pixel*" [3]. The real innovative idea was to use the best part of the CCD technology, the almost-noiseless charge transfer, and combine it with CMOS. The correlated double sampling (CDS) and the fixed-pattern noise reduction implemented in this first sensor meant that, despite this being the first attempt to use CMOS technology in imaging, the sensor showed some good performance. It is interesting to note that in almost all modern CMOS image sensors, we can find some kind of similarities with the first sensors that were developed almost 25 years ago.

Compared to pure CCDs, CMOSs still had a long way to go, and their application was then limited to simple consumer products like webcams. In 1995, a second CMOS imager was presented by the same group [4]. The on-chip digital interface allowed parallel loading of configuration sequences to control the region of interest (RoI) readout and the integration time. Significantly, the paper concluded mentioning the future development of an on-chip analog-to-digital converter (ADC) for a fully digital interface. In the same year, a second publication [5] showed another example of CMOS/CCD integration, with the clear intent of overcoming the limitations of CMOS-only imagers in terms of sensitivity to blue light and the dark current.

The development of CMOS technology helped in narrowing the performance gap with CCD imagers. This was also aided by the evolution of the consumer market with the ever-increasing demand for low-cost, low-power system-on-chip (SoC) cameras.

10.1.1 MONOLITHIC CMOS

The constant improvement of CMOS imaging technology not only revolutionized the consumer market but represented also a step change in radiation detection instrumentation. Two different kinds of devices were normally used for high-energy physics (HEP) detectors: (1) hybrid active pixel sensors and (2) CCDs [6].

Without going too much into the details, hybrid active pixel sensors consist of two layers: the first is a thick (normally 300-μm), high-resistivity silicon substrate, and the second is a CMOS readout application-specific integrated circuit (ASIC) that is connected to the first via bump-bonding. The hybrid solution is expensive because of the two layers and requires more power. CCDs are less power hungry and expensive, but the slow readout is a limitation in HEP experiments with events on a nanosecond scale.

At the beginning of the twenty-first century, CMOS imagers were successfully demonstrated as detectors [6–12] for charged particles. The monolithic approach combined with low power, relative low cost, and the possibility of complicated on-chip electronics was a key factor in determining this success. Recent advances in CMOS fabrication, like stitching [13–16], opened the door to an even more widespread use of CMOS imagers in large scientific experiments, thanks to the fabrication of devices with areas up to a wafer-scale size [17–24].

In this chapter, we want to give an overview not only of the possibilities but also of the limitations and the trade-offs involving CMOS imagers in the field of radiation detection. Far from being complete and exhaustive, Sections 10.2 and 10.3 will describe first how radiation is detected in silicon and then move to the effects of such radiation on CMOS devices. In Section 10.4, the main sources of noise in CMOS image sensors will be summarized. Section 10.5 will present a typical CMOS image sensor architecture, including some of the most common pixel topologies.

Finally, Section 10.6 is dedicated to the successful examples of CMOS sensors in scientific applications.

10.2 RADIATION IN SILICON

In principle, measuring radiation is a relatively simple process during which the radiation interacts with some material, and any resulting variation in such material or in the overall measuring system is then measured. CMOS sensors are fabricated onto a silicon substrate, with a cross section like the one in Figure 10.1.

Any incident radiation interacts with the silicon substrate, often resulting in electric charge being generated. Measuring the generated charge is effectively measuring the incident radiation. There are many publications and books [25–29] in the literature presenting a detailed analysis of radiation–matter interaction. In this chapter, we will present only a brief overview.

When discussing the possible interactions of radiation with matter, the first distinction to make is between charged particles and neutral particles like photons.

Photon (or γ-ray) interaction with matter normally results in three different kinds of effects:

1. Photoelectric absorption
2. Compton scattering
3. Pair production

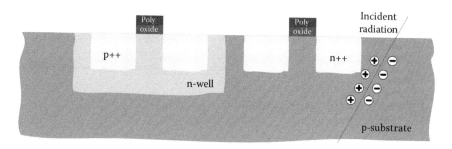

FIGURE 10.1 CMOS process cross section.

In the photoelectric absorption process, the incident photon completely disappears, generating a photoelectron and sometimes x-ray photons. The Compton scattering process occurs when an incident photon loses some of its energy resulting in a scattered electron and a scattered photon of different energy.

In case of γ-rays with energy above several megaelectronvolts, we can have electron–positron (pair) production. Because of the annihilation of the positron, photons are a secondary product of this process.

Like γ-rays, neutrons' interaction with matter is not affected by Coulomb force, so they can travel some considerable distance before undergoing any kind of interaction. There are many possible interaction mechanisms for neutrons, and their description is beyond the scope of this chapter. The detection of neutrons is normally achieved by converting them into secondary charged particles [28].

Unlike photons or neutrons, charged particles are subject to Coulomb force from the electrons and the nuclei of the material that they interact with, and in the large majority of cases, the nuclear interactions are negligible compared to the electric ones. Such interactions are quite complicated to describe and analyze in detail. To describe the behavior of a charged particle, the concept of stopping power is introduced. The stopping power represents the rate of energy loss of a particle, and it is described by the Bethe formula [27,28]:

$$-\frac{dE}{dx} = \frac{4\pi e^4 z^2}{m_0 v^2} NZ \left[\ln \frac{2m_0 v^2}{I} - \ln \left(1 - \frac{v^2}{c^2} \right) - \frac{v^2}{c^2} \right] \tag{10.1}$$

where v and ze are the velocity and the charge of the incident particle, Z is the atomic number of the absorbing material, m_0 is the electron rest mass, e is the electron charge, and I is the ionization potential of the medium [27,28].

Electron interactions differ from heavily charged particles mainly because of the difference in mass. The energy loss is not just due to electronic interaction but also to radiative processes, referred to with the term *bremsstrahlung*, which literally means *breaking radiation*.

The specific energy loss for *bremsstrahlung* can also be calculated [27,29]. It has to be noted that for high-energy electrons, *bremsstrahlung* is the dominant effect [28].

10.2.1 CHARGE COLLECTION

In Section 10.2, we have seen that when radiation interacts with a semiconductor, charge is generated. More generally, we can identify three different phenomena:

1. Excitation of the lattice, when the incident radiation deposits energy to the lattice, generating lattice vibrations
2. Electron–hole pair (EHP) production or ionization
3. Atom displacement

It is obvious that ionization is the process that we are most interested in. Because of the periodic lattice of crystalline materials, electrons can occupy only specific energy levels. One electron can move from the valence band to the conduction band only if it gains sufficient energy, corresponding to the bandgap (separation between the bands) energy. Such a process results not only in an electron in the conduction band but also in a vacancy in the valence band. The two phenomena are collectively known as EHPs.

Semiconductors like silicon are known to be characterized by a distance between the valence band and the conduction band of a few electronvolts (Figure 10.2).

The number of electrons (charge) generated can be expressed as

$$Ne = \Delta E / E_g \qquad (10.2)$$

where ΔE is the energy loss, and E_g is the energy required for an electron to move from the valence band to the conduction band.

Experiments have shown that the number of EHPs detected is much lower than the one that is predicted by Equation 10.2. This happens because the energy lost by the incident radiation in the semiconductor material will indeed generate some charge (EHP) but will also be lost to the lattice. The energy lost to the lattice is accounted for by introducing the so-called Fano factor [27–29]. Such factor value always lies between 0 and 1, and it is equal to 0.115 in silicon.

When the number of electrons and holes in the silicon are the same, the semiconductor is defined as intrinsic. Charge concentration in this case depends only on the temperature and the effective density of states in the conduction band N_C and in the valence band N_V [25].

$$n_i = \sqrt{N_C N_V} \, e^{-E_g / 2KT} \qquad (10.3)$$

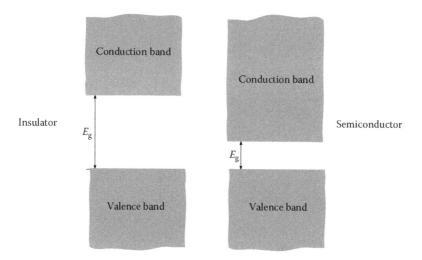

FIGURE 10.2 Insulator and semiconductor energy bands.

High temperature means a higher probability of an electron getting enough energy to jump from the valence band to the conduction band.

In first approximation, we can assume that N_A, the concentration of the acceptor impurities, is equal to N_V and N_D; the concentration of donor impurities is equal to N_C, so Equation 10.3 can be rewritten as

$$n_i = \sqrt{N_D N_A}\, e^{-E_g/2KT} \tag{10.4}$$

The charge carrier density can be changed by adding impurities to the silicon crystalline structure through a process called doping. The total charge density and the resistivity will be [25,27,28]

$$N_{EFF} = N_D - N_A \tag{10.5}$$

$$\rho = \frac{1}{q\left(\mu_H N_A + \mu_E N_D\right)} \tag{10.6}$$

where q is the electron charge, and μ_H and μ_E are the mobility of holes and electrons, respectively.

To understand the basic principles behind charge collection in CMOS sensors, we will start from the well-known p-n junction (Figure 10.3), formed by two adjacent volumes of silicon: one with a high concentration of electrons ($N_D \gg$ ni, n-doped silicon) and one with a high concentration of holes ($N_A \gg$ ni, p-doped silicon).

Because of the imbalance in the electron and hole densities when the p and n volumes are put in contact, there will be a flow of electrons toward the p-doped volume and a flow of holes in the opposite direction. This will create, in first approximation, a volume that is depleted of charge (depletion region) where an electric field prevents any further flow of charge.

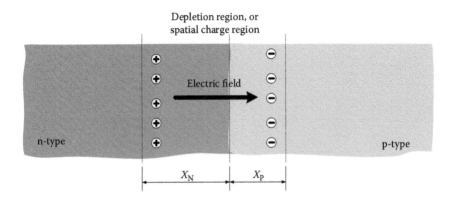

FIGURE 10.3   p-n junction.

The depletion region extends over a distance d, which is given by Equation 10.7, where V is the reverse voltage that is applied to the p-n junction [25,27,28]:

$$d = x_N + x_P = \sqrt{\frac{2\varepsilon V}{q}\left(\frac{1}{N_A} + \frac{1}{N_D}\right)} \tag{10.7}$$

In radiation detection, the depletion region plays a very important role. Electrons and holes generated in this volume will be attracted by the electric field toward different ends, generating a current that is the direct result of radiation.

The depletion region is relatively small, but applying an external potential (reverse biasing of the p-n junction), such volume can be increased and used as an effective radiation detector:

$$\text{If } N_A \gg N_D \quad d = \sqrt{\frac{2\varepsilon V}{q}\frac{1}{N_D}} = \sqrt{2\varepsilon V \rho \mu_H} \tag{10.8}$$

$$\text{If } N_D \gg N_A \quad d = \sqrt{\frac{2\varepsilon V}{q}\frac{1}{N_A}} = \sqrt{2\varepsilon V \rho \mu_E} \tag{10.9}$$

Setting the dopant concentrations and the reverse-biasing V at the right values allows full depletion of the silicon substrate. Charge generated in the depletion region will drift under the effect of the electric field, whereas the current generated in the n and p volumes will diffuse.

From Equations 10.8 and 10.9, it is clear that for a given potential V, the depletion region volume can be increased using high-resistivity silicon substrates also [27,28].

10.3   RADIATION HARDNESS

CMOS sensors can be successfully used as radiation detectors. As mentioned in Sections 10.1 and 10.2, they represent a flexible and cost-effective alternative.

Radiation interaction with a CMOS sensor can result in damages to the sensor itself. Such damages can affect the sensor characteristics, and, if we want to use CMOS in radiation detection, we must be aware of the possible side effects.

10.3.1   Effects of Radiation on CMOS

Sensor–radiation interaction is a complex phenomenon [28]. The outcome is affected by many variables, like the kind of particle (e.g., electron, proton) and its energy, for example. We can distinguish two separated mechanisms: (1) ionization and (2) displacement, the latter often referred to as nonionizing energy loss.

10.3.1.1   Ionization Effects

Incident radiation creates charge not only in the silicon bulk but also in the silicon oxide. In SiO_2, electron mobility is higher than hole mobility, so they move toward the most positive side. Holes, with their lower mobility, have a much higher chance of getting trapped in silicon oxide. Moreover, a hole that reaches the $Si-SiO_2$ interface can be captured by interface traps. This has a direct effect on the behavior of the metal–oxide semiconductor (MOS) transistor.

During the normal operation of an n-type metal–oxide semiconductor (nMOS) transistor, a positive potential is applied to the gate terminal in order to establish a conductive channel (Figure 10.4).

The radiation-induced holes in the oxide will be pushed toward the $Si-SiO_2$ interface by this potential, leading to the buildup of a positively charged layer at the said interface. Such positive charge has to be compensated by a higher gate potential, leading effectively to a shift in the threshold voltage of the MOS transistor [25,27,28].

The charge generated in the oxide is proportional to the oxide thickness, so it is easy to understand that such a problem is exacerbated where the thick field oxide is located. Ionization effects on MOS transistors depend on a variety of factors. We have already mentioned the charge that is trapped in the oxide and at the interfaces and their mobility.

The variations in the time of the electric field have also to be considered, together with the total dose and dose rate. This is why it is best practice to test the radiation tolerance of a CMOS sensor under its normal operating conditions.

10.3.1.2 Displacement Effects

Any incident particle with energy above a few tens of electronvolts has the potential to dislodge one atom of the lattice and create a defect. These kinds of defects can be single-point defects or clusters.

When a lattice atom is displaced and a defect is created, consequently, some midstates appear in the energy band diagram (Figure 10.5).

These midstates can cause carrier recombination and carrier emission, and the net effect depends on many factors like the localization of the midstates and the state of the area (depletion or forward bias). If the area where the defects are created is depleted, such defects will contribute to the reverse current. In case of a forward-biased area, the defects will cause mainly recombination determining a charge loss.

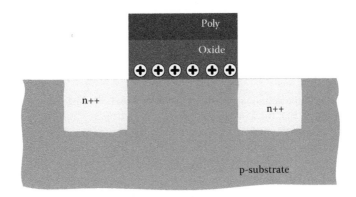

FIGURE 10.4 nMOS transistor cross section with charge in the oxide.

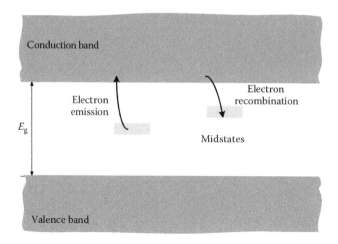

FIGURE 10.5 Band diagram with midstates.

When defects are created close to the conduction boundary, we can see a slightly different effect named *trapping*. Carriers from the conduction band are in fact trapped in these defects and released after some time.

Displacement of atoms can also lead to type inversion in silicon. To better understand this phenomenon, we will focus our attention on a p-n junction that is fully depleted. Equations 10.8 and 10.9 apply to this case. Assuming a starting material doped n with $N_{EFF} = N_D - N_A$, experimental results [26–28,30] show that N_A tends to increase with irradiation resulting in a lower N_{EFF}. Lower effective doping means that the potential required to deplete becomes higher. If the effective doping is inverted, the potential required for depletion might become too high, and effective radiation detection becomes impossible.

10.3.2 RADIATION HARDENING BY DESIGN

Ionization and displacement effects lead to increased leakage current, transistor threshold shift, loss of transconductance, increased leakage current, and increase in noise, especially the $1/f$ [27–30], single event upset (SEU), and single event latchup (SEL), just to mention a few. More generally, we have a degradation of the sensor performance. The constant scaling down of the CMOS process can definitely help (e.g., thinner oxides are less prone to radiation damage), but, at the same time, sensor designers have developed a series of design techniques to prevent or limit the unwanted effects of radiation in CMOS sensors.

Leakage current can be limited by adding substrate contacts whenever possible, bearing in mind that this will impact on the overall integrated circuit area.

Charge generated in the silicon substrate is collected using a diode, whose characteristics, like leakage current, can be affected by radiation too.

For n-well to p-substrate diodes, one of the main issues is represented by the surface component of the total leakage current. The increase of such a current can

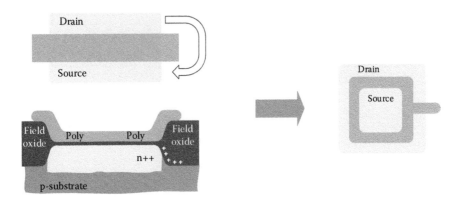

FIGURE 10.6   Radiation-tolerant layout of an MOS transistor.

be mitigated by changing the diode layout and avoiding thick oxide at the diode edges. This can be achieved with the so-called gated diodes [31–33]. Better radiation tolerance can also be obtained with different diode structures [33], controlling the separation of the shallow trench isolation from the edge of the photodiode (PD) itself.

A similar approach cannot be used for different kinds of PDs, like the pinned photodiode (PPD). The PPD [34] has a different structure because of the presence of the p+ *pinning* layer, thanks to which the PD can be fully depleted.

PPDs are widely used in the so-called 4T pixels, described in Section 10.5.1.2. Improvements to the radiation tolerance of the PPDs are not enough to ensure the performance of the overall pixel. More details will be discussed in Section 10.5.1.2.

For metal–oxide-semiconductor field-effect transistor devices, the buildup of positive charge at the Si-SiO_2 interface will cause an increase in leakage current and potentially a short circuit between the source and the drain. This can be prevented with a different gate shape [35–39], as in Figure 10.6.

The resulting MOS is referred to as an enclosed layout transistor (ELT). This is a common technique [36–39] to increase the radiation tolerance of CMOS circuits and has been used with success in many applications. Also, in this case, the price to pay is the silicon area, with the added limitation of a minimum transistor width. Less-known layout solutions to increase the radiation tolerance of the MOS transistor can be found in the literature [40,41].

10.4   NOISE SOURCES

In Section 10.1, we presented a short summary of the rise of CMOS technology and how CCDs were gradually replaced in a large number of applications. We have also compared the monolithic CMOS image sensors with the hybrid detectors in HEP experiments. In both cases, one of the key parameters was noise. The CMOS cannot be a valid alternative if the noise performance is not competitive. This is of particular importance for scientific applications.

In Sections 10.4.1 through 10.4.4, we will present a brief overview of the noise sources in CMOS image sensors.

10.4.1 SHOT NOISE

The core of a pixel in a CMOS imager is represented by a reverse-biased PD. The charge generated by incident radiation will originate what is known as photocurrent. When there is no radiation, the photocurrent is not zero, and it is referred to as dark current.

For a sensor exposed to a uniform flux, the arrival time of each incident particle is a random event that follows a Poisson statistic. This means that the associated noise is proportional to the square root of the mean number of incident particles:

$$I_N = \sqrt{2*I*q*BW}$$

(10.10)

where I is the current, q is the electron charge, and BW is the bandwidth. This is valid for photocurrent shot noise and dark current shot noise. Because shot noise is associated with any current flowing, for the photocurrent and the dark current, we can also write that the noise in electrons is proportional to the square root of the number of electrons:

$$n_{e,ph} = \sqrt{N_{e,ph}},$$

(10.11)

$$n_{e,dc} = \sqrt{N_{e,dc}}$$

(10.12)

The photocurrent and the dark current voltage noise can be calculated from Equations 10.11 and 10.12, respectively, by multiplying by the conversion gain C_G (μV/e); then, using Equation 10.10, we can write

$$\overline{vn_{ph}} = C_G n_{e,ph} = C_G \sqrt{N_{e,ph}} = C_G \sqrt{\frac{I_{ph} T_{int}}{q}}$$

(10.13)

$$\overline{vn_{dc}} = C_G n_{e,dc} = C_G \sqrt{N_{e,dc}} = C_G \sqrt{\frac{I_{dc} T_{int}}{q}}$$

(10.14)

10.4.2 THERMAL NOISE

Analysis of MOS transistors' noise plays a crucial role in understanding noise sources and possible mitigation strategies in CMOS imagers. The noise sources of an MOS transistor are reported in Figure 10.7 [42].

The current source on the left, i_g^2, represents the noise that is associated with the gate leakage current. For the current source on the right, i_d^2, it is well known that we

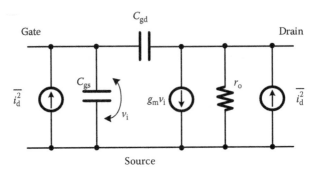

FIGURE 10.7 MOS transistor model completed with noise sources.

can identify two main noise contributions: (1) thermal noise and (2) flicker or $1/f$ noise.

The random motion of charge carrier in conductors is the reason behind the thermal or white noise. A simple resistor has a noise of

$$v_n^2 = 4kTR * BW, \quad i_n^2 = 4kT\frac{1}{R} * BW \tag{10.15}$$

For the MOS transistor, the thermal contribution to noise source i_d^2 in Figure 10.7 is equal to [42]

$$\overline{i_{d,th}^2} = 4kT\left(\frac{2}{3}g_m\right)BW \tag{10.16}$$

It is worth noticing that reducing the bandwidth reduces the noise or, in other words, that the bandwidth is a factor that determines the total noise. A well-known example of this is the so-called reset noise or KTC noise, one of the dominant sources of noise in CMOS image sensors. If we consider the simple circuit in Figure 10.8, we can say that the MOS transistor operates as a switch and when closed can be approximated as a resistance.

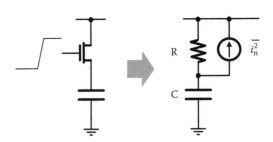

FIGURE 10.8 Reset noise.

The noise generated by a resistance R (see Equation 10.15) has to be multiplied by the circuit bandwidth and integrated over the full-frequency range. Without showing all the math behind, the result is that the voltage noise due to the resetting of the capacitance voltage is KT/C, with C being the capacitance under reset.

To calculate such noise as a charge, we multiply it by C^2 (that's why KTC noise) and then divide it by 1.6×10^{-19} (charge on single electron) to get the noise in electrons.

10.4.3 1/F NOISE

Flicker noise can be explained as the random current variation because of carriers' trapping and detrapping by defects at the $Si\text{-}SiO_2$ interface. According to this model, known as the McWhorter model, flicker noise is a surface effect. A second model by Hooge links the flicker noise to the random variation in carrier mobility. We are not entering the debate here, and we are just reminding the readers of the flicker component of the i_d^2 from Figure 10.7 [42,43]:

$$\overline{i_{d,fn}^2} = \frac{KI_D}{C_{OX}L^2 f}$$

(10.17)

where K is a process-dependent parameter, I_D is the source drain current, C_{OX} is the oxide capacitance per unit area, L is the channel length, and f is the frequency.

10.4.4 RANDOM TELEGRAPH SIGNAL NOISE

A special case of $1/f$ noise is represented by the random telegraph signal (RTS) noise. According to the McWhorter model, the larger the number of trapping centers, the wider the current fluctuations and hence the noise. In a deep submicron process, it is not uncommon to have a situation where a transistor has just one trapping center; hence, the flicker noise will result in a random variation between two levels.

The net result is that for small-size transistors, like the in-pixel ones, the noise spectrum might look more like an RTS noise than a $1/f$ noise. This is something that has to be taken into consideration when designing or testing a CMOS sensor [44,45].

10.5 CMOS SENSOR ARCHITECTURE

In Section 10.1.1, we did mention that one of the advantages of CMOS technology in imaging and in radiation detection is the monolithic approach. Readout electronics and the sensitive area are together on the same silicon support.

The light/radiation-sensitive area is obviously the pixel array. On the same ASIC, we have other electronic blocks that make possible the sensor operation and the readout of the information that is gathered by the pixel array. A block diagram of a CMOS image sensor is shown in Figure 10.9.

The sensor operates by collecting in the pixel array the charge that is generated by the incident radiation. Such information is then transferred to the *sampling and storing stages*. A second stage is normally added to read out such information in the most convenient way. If the sensor operates on a row-by-row basis, we have the

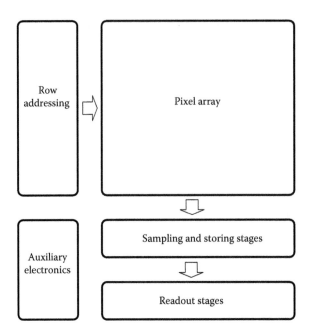

FIGURE 10.9   Block diagram of a CMOS image sensor.

so-called rolling shutter; if the pixel array data are read out all at the same time, we refer to this as a global shutter. The row or rows to be read out are selected by the *row-addressing* block.

Information in the *readout stages* can be shifted out of the sensor in analog or digital format.

10.5.1   PIXELS

The pixel array is the core of CMOS sensor. The charge generated by the incident radiation is converted into a voltage or a current that can be easily processed by the readout electronic. Since the first sensor [3,4], a lot of effort has gone into the design of pixels. Sections 10.5.1.1 through 10.5.1.3 will present some of the most common topologies, with particular attention to their application to the detection of radiation for scientific applications.

10.5.1.1   3T

A p-n junction can be used as the simplest of the detectors. The electric field of the spatial charge area forces the drift of the charge that is generated, resulting in a photocurrent. The simplest readout architecture for such current is the so-called 3T pixel, shown in Figure 10.10.

The operation of such pixel is very well known and well documented in the literature ([3,6,10,11,15,16,19,24,29], just to mention publications that were already cited). We would like to focus the readers' attention on some important characteristics of this pixel topology. First of all, the p-n junction or diode does not cover the entire

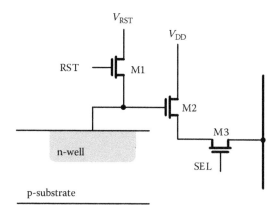

FIGURE 10.10 3T pixel.

area of the pixel; hence, the charge generated is collected not only by drift but also by diffusion.

The p-n junction is the only n-well in the pixel; the other active components are all nMOS. In this way, the charge diffusing in the p-substrate can be collected only by the diode or recombination.

The third observation is relative to the radiation tolerance; an ELT design for the nMOS and a gated diode can considerably improve the radiation resistance of the 3T pixel.

10.5.1.2 4T

The 3T pixel performance is limited by the well-known reset noise or KTC noise. The absence of an in-pixel memory makes CDS not possible. In 1982, N. Teranishi and his research group at NEC [46] presented what they refer to as a *no-image-lag photodiode*. It was the first example of what is now known as a pinned photodiode.

Initially applied to interline CCD sensors, the PPD was combined with CMOS technology in the 1990s [5,46] originating the 4T pixel, depicted in Figure 10.11. The term 4T is derived from the presence of a fourth transistor: the transfer gate (TG).

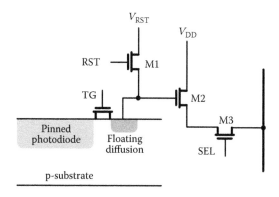

FIGURE 10.11 4T pixel.

CDS gives to 4T pixels a clear advantage in terms of noise performance. Similarly to 3T pixel, the 4T one is made of nMOS only.

As a radiation detector, the 4T pixel suffers from radiation damage [47]. Among the different effects reported in the literature [47–49], we mention the increase of the pinning voltage with total ionizing dose (TID), the decrease of the PD full-well capacity, worse charge transfer efficiency, and noise increase [47–49]. As discussed in Section 10.3.2, acting on the diode design is normally not enough to improve the radiation tolerance. The shape of the TG also plays an important role, as recently demonstrated on a 180-nm process [49].

10.5.1.3   In-Pixel Processing

The 3T and 4T pixel topologies shown in Sections 10.5.1.1 and 10.5.1.2 are a good choice for a large number of different applications, but high-end scientific projects often demand more. Pixel topologies are constantly evolving, with more functionalities and more active components that are added to the pixel itself, like the 5T pixels for global shutter readout.

Sensors for the detection of charged particles in HEP experiments like the MIMOSA-26 [50–52] and MIMOSA-32 [53] or sensors for mass spectroscopy like PIMMS [54,55] are the perfect examples of pixels with added functionalities.

Complex functions in pixels cannot be implemented without using p-type metal–oxide semiconductor (pMOS) transistors. This may be problematic because the n-well associated with such transistors will act like a diode, effectively *stealing* signals from the pixel diode.

This problem has been solved by developing a special technology with an implant that can shield the n-wells for pMOS transistor from the silicon substrate. In Figure 10.12, the cross section of the InMAPS process [55–59] is presented.

Charge generated by the incident radiation is collected by the diode (n-well on the right in Figure 10.12) and not by the n-well of the pMOS transistors. The presence of nonstandard (for CMOS processes) implants like the deep p-well made it possible to design analog conditioning circuits and digital logic in each PIMMS pixel [54,55] using more than 600 transistors per pixel.

The promising results obtained are an indication that the way toward more and more *in-pixel intelligence* will be possible in the future.

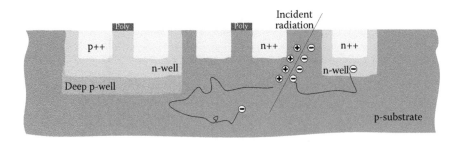

FIGURE 10.12   InMAPS process cross section.

10.5.2 Row Addressing

Rolling shutter or global shutter architecture both need a way to select the pixels to be read out. We can distinguish three main ways of achieving this task:

1. External addresses and on-chip decoding
2. Shift register-based row addressing
3. On-chip addressing and decoding

The first case, shown in Figure 10.13, is the simplest one, requiring very limited on-chip logic. It is relatively easy to build redundancy to maximize yield; there are no major limitations to the readout so that the RoI can be implemented, and the approach can be easily scaled to large-area devices. The main drawback is the number of external connections and the complexity of the camera system.

Replacing the decoding stage with a shift register (see Figure 10.14) simplifies the sensor interface, but will introduce some limitations in terms of speed, especially for large devices. It is harder to build redundancy in the system, and the failure of one stage could compromise the entire sensor. RoI readout can be implemented, and these limitations can be overcome by increasing the complexity of the shift register.

The flexibility and robustness of the decoder approach and the simple interface of the shift register solution can be achieved by on-chip addressing generation and decoding, shown in Figure 10.15.

The external bus for the row addresses is replaced by an on-chip address generator, like a counter with programmable start and stop values.

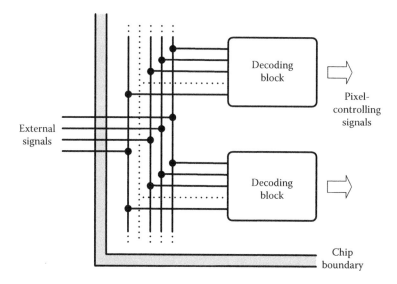

FIGURE 10.13 External addresses and on-chip decoding.

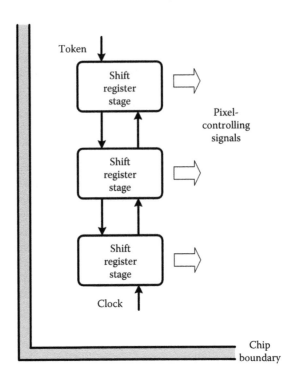

FIGURE 10.14   Shift register-based row addressing.

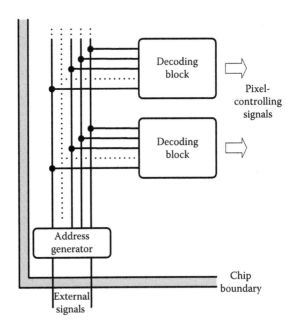

FIGURE 10.15   On-chip addressing and decoding.

Readout speed can be improved by having reset and select signal in parallel, applied to different rows, so that when row N is read out, row $N - 1$ is reset (Figure 10.16).

Further increase in the readout speed can be obtained by reading several rows at the same time (see Figure 10.17). This has the drawback of complicating the readout circuits, requiring more power and complicating the pixel layout. Moreover, the added lines for parallel readout will affect the sensor QE.

Increasing the readout speed is of particular importance for large-area sensors, where the frame rate can be affected by the delays due to longer connection lines and the large number of pixels to be read out. PERCIVAL [60–64], a large-area CMOS sensor for low-energy x-rays, has a seven-row readout architecture.

10.5.3 SENSOR READOUT

In Section 10.5.2, we have seen how we can select which pixels we want to read out. The readout operation can be seen as a multiplexing operation. Depending on the

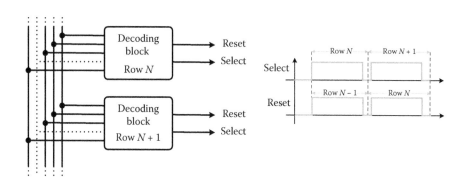

FIGURE 10.16   Row reset and select.

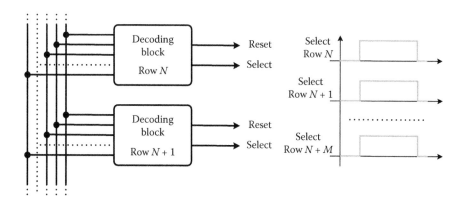

FIGURE 10.17   Multiple rows addressing.

pixel data format and on the amount of on-chip processing, we can have three different cases:

1. Analog pixel data and analog output
2. Digital pixel data and digital output
3. Analog pixel data and digital output

In the first case, the information from the pixel array is transferred outside the pixel array, normally stored on capacitors. The M columns are then multiplexed to the N outputs, normally with the help of dedicated on-chip electronics. This is the architecture that was used in the first CMOS sensor, and it is still quite common.

The advantage of this approach is a relatively simple architecture, but there are some limitations in terms of speed, noise, and complexity of the external acquisition system.

Noise problems, speed limitations, and complicated DAQ (Data AcQuisition) system issues can be mitigated in the digital readout of digital pixel data. The entire readout chain is digital, so it is suitable for high-speed readout.

The fact that each pixel has its own ADC might look as a great advantage, but a series of constraints and limitations have to be considered. First of all is the pixel area; small pixels will not be possible, in particular, considering the fact that radiation-tolerant designs require larger areas. Power consumption and complexity have to be considered too. Power distribution has to be carefully considered, and a similar consideration can be applied to signal distribution. A clock distribution tree, spread all over the pixel array, for example, is likely to have some kind of impact on the noise performance.

Moreover, in-pixel ADCs are likely to require more controlling signal, so a reduction in fill factor has to be considered too.

The third readout architecture is nowadays the most common. Data from the pixel array are converted into a digital format by on-chip ADCs, often column-parallel ADCs. Many different architectures can be used for the column-parallel ADCs, with single ramp and successive approximation register (SAR) being the most common ones.

Recent publications [65–67] have shown that sigma–delta ADCs are also a very good solution for column-parallel ADCs, and a separate chapter of this book is dedicated to this.

The constant trend toward faster sensors with more and more pixels is pushing the output data rate toward the gigapixels per second. Larger sensors with multimegapixel arrays and higher frame rates mean that high-speed digital outputs like current mode logic (CML) or low voltage differential signaling (LVDS) are becoming a necessity in state-of-the-art radiation detectors.

10.6 TRENDS AND APPLICATIONS

CMOS image sensors are becoming more and more common in the detection of charged particles. Radiation hardness and a good signal/noise ratio are very important characteristics to be considered.

Section 10.3.1.2 presented a summary of the damages to silicon due to radiation. In particular, we discussed the *bulk damages* and the associated signal loss. We have also presented the benefit of charge collection by drift in the depleted region of a p-n junction. Fully depleted detectors to combat radiation damages are particularly important in HEP applications. One way to achieve a fully depleted detector is to exploit the characteristics of the high-voltage complementary metal–oxide semiconductor (HVCMOS) processes that are normally used for automotive applications (Figure 10.18) [68–74].

Using the HVCMOS process [68–74], it is possible to design the entire pixel inside the deep n-well that is used to collect the charge. The main advantages of this are increased radiation tolerance to nonionizing damage and the possibility to have in-pixel electronics. The drawbacks are represented by a larger capacitance of the collecting electrode and noise performance. Nevertheless, this shows that CMOS technology can be applied with success to applications in radiation-harsh environments.

Synchrotron radiation sources' brilliance is constantly increasing, so new detectors with high frame rate, low noise (single photon capability), and large dynamic rage are always in demand. The CMOS monolithic approach can represent a valid solution in these cases, in particular, if we consider soft x-ray applications with an energy range from 250 to 1–2 keV. High efficiency and low noise can be achieved with other technologies, but if higher frame rates are required, then the advantage represented by monolithic CMOS technology is clear. This is exactly what the PERCIVAL project, a collaboration between Deutsches Elektronen-Synchrotron (DESY), STFC-RAL, Elettra, diamond light source (DLS), and Pohang, has proven [60–64]. The combination of back-thinning and CMOS technology made it possible to design and test a prototype that is suitable for the energy range that was mentioned above, together with a frame rate of 120 fps and a dynamic range of 10^5.

Particle physics is probably the most obvious field of application where CMOS image sensors can make a difference, but not the only one. Over the years, we have witnessed an extraordinary change in the transmission electron microscopy (TEM) field. In microscopy, the limitations to the resolution that can be achieved with light (photons) can be overcome by choosing different particles like electrons. To a first degree of approximation, the resolution that can be obtained with light microscopes is limited by the wavelength of the light; ultraviolet (UV) microscopes can increase the resolution up to a factor of 2. Wave–particle duality theory means that one electron can behave like a particle and like a wave, with a wavelength that is given by the De Broglie equation. The specimen under investigation is irradiated by a beam of electrons, and the resulting image contains details that were impossible to see with photons.

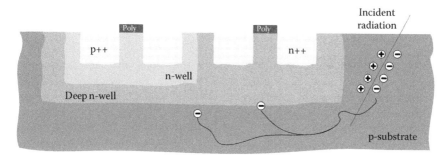

FIGURE 10.18 HVCMOS sensor cross section.

Such technology was developed starting from 1928, and the first commercial transmission electron microscope was delivered in 1939.

For a long time, the image generated by the electron beam has been recorded using photographic film. It is easy to imagine that, even if the resolution and the signal/noise ratio were sufficient to deliver good-quality images, the overall process must have been really slow. Film was replaced in the 1990s by a CCD camera that is coupled with a fiber optic phosphor (FOP). CCD is a digital imaging device, so there is no more need for lengthy film processing. Unfortunately, such advantage had to be traded off with the fact that CCDs with FOP are not suitable for direct detection. The sensor was not in fact in the electron beam, and the FOP leads to considerable loss of performance. This because of the poor radiation tolerance that is typical of CCD sensors.

CMOS image sensors were able to combine the direct detection that is possible with film and the fast and digital support that is represented by the CCDs. This is a perfect example of how disruptive CMOS sensors can be in scientific applications.

The first commercially available transmission electron microscope to have a CMOS sensor was the FEI FALCON. The sensor in the microscope was developed by a consortium that was led by STFC-RAL and LMB-MRC [75–82]. It is a 16-million pixel sensor with analog outputs. All the electronic blocks have been designed to be rad-hard, achieving an overall tolerance in excess of 20 Mrad.

CMOS technology combined with cryo-EM represents a step change in microscopy, and the recent literature is a testament to such revolution [83–86]. Such success is confirmed by the fact that all the major players in the TEM field now have CMOS-based cameras [87–89].

In this chapter, we have mentioned several times the progress that was made by CMOS technology as a radiation detector. It seems almost natural that, recently, a novel radiation monitor, a compact instrument named a highly miniaturized radiation monitor, has been developed [90–92] and launched into space [93]. The instrument is based on a CMOS image sensor that was designed in commercial 0.18-µm process with 4T pixels. The 4T pixel has been chosen to provide the low noise that is required, and despite not being normally suitable for radiation-harsh environments, such a pixel has been proven to be usable up to a few hundreds of radians [47]. CMOS imaging technology made possible improvements that are measured in the order of magnitude for what concerns size, weight, and power consumption.

10.7 CONCLUSIONS

CMOS sensors for scientific applications are far from being an established technology, and, in many cases, they are not the first choice. But, we cannot ignore the remarkable achievements of the past years, some of which have been presented in this chapter.

CMOS has the potential to be a truly disruptive technology in the development of scientific instruments. The constant evolution of the technology, and the relatively easy access to it, coupled with the efforts of many scientists and researchers around the world, is almost a guarantee of more to come.

REFERENCES

1. P. Rousseau. 2009. CMOS: An emerging technology system driver. In *Circuit at the Nanoscale*, edited by Kris Inieswki, CRC Press, Boca Raton, FL, pp. 3–9.
2. W. Boyle and G. Smith. 1970. Charge coupled semiconductor devices. *Bell System Technical Journal*, vol. 49, pp. 587–593.
3. E. Fossum et al. 1993. A 128 × 128 CMOS active pixel image sensor for highly integrated imaging systems. *IEEE IEDM Technical Digest*, pp. 583–586.
4. E. Fossum et al. 1995. 128 × 128 CMOS photodiode-type active pixel sensor with on-chip timing, control and signal chain electronics. *Charge-Coupled Devices and Solid-State Optical Sensors V, Proceedings of SPIE*, vol. 2415, pp. 117–123.
5. E. Fossum et al. 1995. An active pixel sensor fabricated using CMOS/CCD process technology. *IEEE Workshop on Charge-Coupled Devices and Advanced Image Sensors*, pp. 115–119, Dana Point, California, April 20–22.
6. R. Turchetta. 2001. CMOS sensors for the detection of minimum ionizing particles. *IEEE Workshop on Charge-Coupled Devices and Advanced Image Sensors*, pp. 7–9, Lake, Tahoe, Nevada, June.
7. G. Deptuch et al. 2000. Simulation and measurements of charge collection in monolithic active pixel sensors. *International Workshop on Semiconductor Pixel Detectors for Particles and X-Rays*, Genova, 5–8 June.
8. G. Claus et al. 2000. Particle tracking using CMOS monolithic active pixel sensor. *International Workshop on Semiconductor Pixel Detectors for Particles and X-Rays*, Genova, 5–8 June.
9. G. Deptuch et al. 2000. Design and testing of monolithic active pixel sensors for charged particle tracking. *IEEE Nuclear Science Symposium and Medical Imaging Conference*, 15–20 October.
10. R. Turchetta et al. 2001. A monolithic active pixel sensor for charged particle tracking and imaging using standard VLSI CMOS technology. *Nuclear Instruments and Methods in Physics Research A*, vol. 458, pp. 677–689.
11. R. Turchetta. 2006. A CMOS monolithic active pixel sensors (MAPS) for future vertex detectors. *SNIC Symposium*, Stanford, CA, 3–6 April.
12. B. Dierickx et al. 1997. Near 100% fill factor CMOS active pixels. *IEEE Charge-Coupled Devices and Advanced Image Sensors Workshop*, Brugge, Belgium, June 5–7.
13. A. Theuwissen et al. 1991. A 2.2 Mpixel FT CCD imager according to Eureka HDTV standard. *IEDM Technical Digest*, pp. 167–170.
14. G. Kreider et al. 1995. An mK × nK modular image sensor design. *IEDM Technical Digest*, pp. 155–158.
15. S. Bohndiek et al. 2009. Characterization and testing of LAS: A prototype "large area sensor" with performance characteristics suitable for medical imaging applications. *IEEE Transactions on Nuclear Science*, vol. 56, no. 5, pp. 2938–2946.
16. A. Clark et al. 2008. A 54 mm × 54 mm–1.8 megapixel CMOS image sensor for medical imaging. *Proceedings of the Nuclear Science Symposium*, pp. 3814–3818, Dresden, Germany, October 19–25.
17. R. Reshef et al. 2009. Large format medical x-ray CMOS image sensor for high resolution high frame rate applications. *Proceedings of the International Image Sensor Workshop*, Bergen, Norway, June 22–28.
18. L. Kourthout et al. 2009. A wafer-scale CMOS APS imager for medical x-ray applications. *Proceedings of the International Image Sensor Workshop*, Bergen, Norway, June 22–28.
19. M. Esposito et al. 2011. DynAMITe: A wafer scale sensor for biomedical applications. *9th International Conference on Position Sensitive Detectors*, Aberystwyth, United Kingdom, September 12–16.

20. H. Takahashi et al. 2011. A 300 mm wafer-size CMOS image sensor for low level light imaging. *Proceedings of the International Image Sensor Workshop*, Hoikkaido, Japan, June 8–11.
21. T. Anaxagoras et al. 2012. Radiation hardness of a large area CMOS active pixel sensor for bio-medical applications. *Proceedings of the Nuclear Science Symposium*, Anaheim, California, October 27–November 3.
22. J. B. Kim. 2015. A high-speed wafer-scale CMOS x-ray detector with column-parallel ADCs using oversampling binning method. *IEEE Transaction on Electron Devices*, vol. 62, no. 3.
23. Available at http://www.forzasilicon.com/2014/06/wafer-scale-x-ray-cmos-image-sensor/.
24. I. Sedgwick et al. 2013. LASSENA: A 6.7 megapixel, 3-sides buttable wafer-scale CMOS sensor using a novel grid-addressing architecture. *Proceedings of the International Image Sensor Workshop*.
25. S. Sze. 1981. *Physics of Semiconductor Devices*. New York: Wiley.
26. A. Holmes-Siedle, L. Adams. 2000. *Handbook of Radiation Effects*. Oxford, UK: Oxford University Press.
27. G. Knoll. 2000. *Radiation Detection and Measurement*. New York: Wiley.
28. S. N. Ahmed. 2007. *Physics and Engineering of Radiation Detection*. San Diego: Academic Press.
29. D. Durini. 2014. *High Performance Silicon Imaging*. Sawston, Cambridge, UK: Woodhead Publishing.
30. H. Spieler. 2005. *Semiconductor Detector Systems*. Oxford, UK: Oxford University Press.
31. A. Grove, D. Fitzgerald. 1966. Surface effects on p-n junctions: Characteristics of surface space-charge regions under non-equilibrium conditions. *Solid-State Electronics*. Oxford, UK: Pergamon Press, vol. 9, pp. 783–806.
32. A. Czerwinski. 2003. Gated-diode study of corner and peripheral leakage current in high-energy neutron irradiated silicon p-n junctions. *IEEE Transactions on Nuclear Science*, vol. 50, no. 2, pp. 278–287, April.
33. V. Goiffon et al. Generic radiation hardened photodiode layouts for deep submicron CMOS image sensor processes. *IEEE Transaction on Nuclear Sciences*, vol. 58, no. 6, pp. 3076–3084, ISSN 0018-9499.
34. E. Fossum, D. Hondongwa. A review of the pinned photodiode for CCD and CMOS image sensors. *IEEE Journal of the Electron Devices Society*, vol. 2, no. 3, pp. 33–43, May.
35. H. L. Hughes, J. M. Benedetto. 2003. Radiation effects and hardening of MOS technology: Devices and circuits. *IEEE Transactions on Nuclear Science*, vol. 50, no. 3, pp. 500–521, June.
36. G. Anelli et al. 1999. Radiation tolerant VLSI circuits in standard deep submicron CMOS technologies for the LHC experiments: Practical design aspects. *IEEE Transactions on Nuclear Science*, vol. 46, no. 6, pp. 1690–1696, December.
37. M. Campbell et al. 1999. A pixel readout chip for 10-30 MRad in standard 0.25pm CMOS. *IEEE Transactions on Nuclear Science*, vol. 46, no. 3, pp. 156–160, June.
38. W. Snoeys et al. 2000. Integrated circuits for particle physics experiments. *IEEE Journal of Solid-State Circuits*, vol. 35, no. 12, pp. 2018–2030, December.
39. W. Snoeys et al. 2002. A new NMOS layout structure for radiation tolerance. *IEEE Transactions on Nuclear Science*, vol. 49, no. 4, pp. 1829–1833, August.
40. S. P. Gimenez et al. 2014. Improving MOSFETs radiation robustness by using the wave layout to boost analog ICs applications. *29th Symposium on Microelectronics Technology and Devices (SBMicro)*, Aracaju, September 1–5.
41. S. P. Gimenez. 2015. An innovative non-standard layout style (diamond) to boost electrical performance and the radiation tolerance of MOSFETs, focussing on space and medical CMOS ICs applications. *Proceedings of CMOS Emerging Technologies Conference*, Vancouver, Canada, May 20–22.

42. P. Gray, P. Hurst, S. Lewis, R. Meyer. 2001. *Analysis and Design of Analog Integrated Circuits*. New York: J. Wiley & Sons Press.
43. P. Allen, D. Holberg. 2002. *CMOS Analog Circuit Design*. Oxford, UK: Oxford University Press.
44. M. Deveaux et al. 2008. Random telegraph signal in monolithic active pixel sensors. *IEEE Nuclear Science Symposium*, Dresden, Germany, October 19–25.
45. V. Goiffon et al. 2014. Influence of transfer gate design and bias on the radiation hardness of pinned photodiode CMOS image sensors. *IEEE Transactions on Nuclear Science*, vol. 61, no. 6, pp. 3290–3301, December.
46. N. Teranishi et al. 1982. No image lag photodiode structure in the interline CCD image sensor. *Proceedings of International Electron Devices Meeting*, vol. 12, no. 6, pp. 324–327.
47. R. Coath et al. 2010. A low noise pixel architecture for scientific CMOS monolithic active pixel sensors. *IEEE Transactions on Nuclear Science*, vol. 57, no. 5, pp. 2490–2496, October.
48. M. Innocent. 2013. A radiation tolerant 4T pixel for space applications. *Proceedings of Workshop on CMOS Image Sensors for High Performance Applications*, Toulouse, France, November.
49. V. Goiffon et al. 2012. Radiation effects in pinned photodiode CMOS image sensors: Pixel performance degradation due to total ionizing dose. *IEEE Transactions on Nuclear Science*, vol. 59, no. 6, pp. 2878–2887.
50. M. Winter et al. 2009. First test results of MIMOSA-26, a fast CMOS sensor with integrated zero suppression and digitized output. *IEEE Nuclear Science Symposium*, pp. 1169–1173, Orlando, USA, October 24–November 1.
51. E. C. Aschenauer et al. 2012. Monolithic active pixel silicon detectors for future electron ion colliders: Status and plans. *Proceedings of the Nuclear Science Symposium*, pp. 1370–1372, Anaheim, USA, October 27–November 3.
52. M. Winter. 2010. From MIMOSA-26 to ULTIMATE. Presentation at the STAR-PXL ULTIMATE Review—BNL, December.
53. H. Hillemanns. 2013. Radiation hardness and detector performance of new 180 nm CMOS MAPS prototype test structures developed for the upgrade of the ALICE Inner Tracking System. *Proceedings of the Nuclear Science Symposium*, Seoul, South Korea, October 27–November 2.
54. J. J. John et al. 2011. PImMS, a fast event-triggered monolithic pixel detector with storage of multiple timestamps. *9th International Conference on Position Sensitive Detectors*, Aberystwyth, United Kingdom, September 12–16.
55. I. Sedgwick et al. 2012. PImMS: A self-triggered, 25 ns resolution monolithic CMOS sensor for time-of-flight and imaging mass spectrometry. *IEEE 10th International New Circuits and Systems Conference*, pp. 497–500, Quebec, Canada, June 17–20.
56. J. Crooks et al. 2007. A novel CMOS monolithic active pixel sensor with analog signal processing and 100% fill factor. *Proceedings of the Nuclear Science Symposium*, pp. 931–935, Honolulu, USA, October 26–November 3.
57. R. Turchetta et al. 2008. TPAC: A 0.18 micron MAPS for digital electromagnetic calorimetry at the ILC. *Proceedings of the Nuclear Science Symposium*, pp. 2224–2227, Dresden, Germany, October 19–25.
58. S. Zucca et al. 2012. Monolithic pixel sensors for fast particle trackers in a quadruple well CMOS technology. *Proceedings of the Nuclear Science Symposium*, pp. 1742–1749, Anaheim, USA, October 27–November 2.
59. L. Ratti et al. 2013. Monolithic pixel sensors for fast silicon vertex trackers in a quadruple well CMOS technology. *IEEE Transaction on Nuclear Science*, vol. 60, no. 3, pp. 2343–2351.
60. C. B. Wunderer et al. 2014. The PERCIVAL soft x-ray imager. *JINST* 9 C03056.

61. C. B. Wunderer et al. 2014. Percival: an international collaboration to develop a MAPS-based soft x-ray imager. *Synchrotron Radiation News* vol. 27, no. 30, pp. 30–34.

62. C. B. Wunderer et al. 2014. The PERCIVAL soft x-ray imager. *Proceedings of 16th International Workshop on Radiation Imaging Detectors*, Trieste, Italy, June 22–26.

63. B. Marsh et al. 2014. PERCIVAL: The design and characterization of a CMOS image sensor for direct detection of low-energy x-rays. *Proceedings of the Nuclear Science Symposium*, Seattle, WA, United States, November 8–15.

64. I. Sedgwick. 2014. The PERCIVAL CMOS image sensor. *Proceedings of Front End Electronics*, United States, May 19–23.

65. A. Xhakoni et al. 2014. A low-noise high-frame-rate 1-D decoding readout architecture for stacked image sensors. *IEEE Sensors Journal*, vol. 14, no. 6, pp. 1966–1974.

66. B. Wang et al. 2013. A 1.8-V 14-bit inverter-based incremental $\Sigma\Delta$ ADC for CMOS image sensor. *Proceedings of the 10th International Conference on Application-Specific Integrated Circuits*, ASICON, Shenzhen, China, October 28–31.

67. Y. Oike, A. El Gamal. 2013. CMOS image sensor with per-column $\Sigma\Delta$ ADC and programmable compressed sensing. *IEEE Journal of Solid-State Circuits*, vol. 48, no. 1, pp. 318–328.

68. I. Peric. 2007. A novel monolithic pixelated particle detector implemented in high-voltage CMOS technology. *Nuclear Instruments and Methods in Physics Research A* vol. 582, pp. 876–885.

69. I. Perić, C. Takacs. 2010. Large monolithic particle pixel-detector in high-voltage CMOS technology. *Nuclear Instruments and Methods in Physics Research A* vol. 624, pp. 504–508.

70. I. Peric, C. Kreidl and P. Fischer. 2011. Particle pixel detectors in high-voltage CMOS technology—New achievements. *Nuclear Instruments and Methods in Physics Research A* vol. 650, no. 1, pp. 158–162.

71. I. Peric et al. 2011. The first beam test of a monolithic particle pixel detector in high-voltage CMOS technology. *Nuclear Instruments and Methods in Physics Research A* vol. 628, pp. 287–291.

72. I. Peric. 2012. Active pixel sensors in high-voltage CMOS technologies for ATLAS. *Workshop on Intelligent Trackers*, Pisa, Italy, May 3–5.

73. N. Berger et al. 2013. A tracker for the Mu3e experiment based on high-voltage monolithic active pixel sensors. *Nuclear Instruments and Methods in Physics Research A* vol. 732, pp. 61–65.

74. I. Peric. 2013. Development of HV CMOS sensors for 3D integration. *Advanced Interconnections for Chip Packaging in Future Detectors*, Frascati, Italy, April 10–12.

75. A. R. Faruqi, R. Henderson. 2007. Electronic detectors for electron microscopy. *Current Opinion in Structural Biology* vol. 17, pp. 549–555.

76. G. McMullan, S. Chen, R. Henderson, A. R. Faruqi. 2009. Detective quantum efficiency of electron area detectors in electron microscopy. *Ultramicroscopy* vol. 109, pp. 1126–1143.

77. G. McMullan, A. R. Faruqi, R. Henderson, N. Guerrini, R. Turchetta, A. Jacobs, G. van Hoften. 2009. Experimental observation of the improvement in MTF from backthinning a CMOS direct electron detector. *Ultramicroscopy* vol. 109, pp. 1144–1147.

78. G. McMullan et al. 2009. Enhanced imaging in low dose electron microscopy using electron counting. *Ultramicroscopy*, vol. 109, no. 12, pp. 1411–1416, November.

79. N. Guerrini et al. 2010. A high frame rate, 16 million pixels, radiation hard CMOS sensor. *Proceedings of 12th International Workshop on Radiation Imaging Detectors*, Cambridge, United Kingdom, July 11–15.

80. G. McMullan et al. 2014. Comparison of optimal performance at 300 keV of three direct electron detectors for use in low dose electron microscopy. *Ultramicroscopy*, vol. 147, pp. 156–163, December.
81. N. Guerrini. 2014. CMOS APS for TEM. *Proceedings of International Workshop on CMOS Active Pixel Sensors for Particle Tracking*, Bonn, Germany, September 1–5.
82. N. Guerrini. 2015. CMOS image sensors for TEM—State of the art and outlook. *Proceedings of CMOS Emerging Technologies Conference*, Vancouver, Canada, May 20–22.
83. W. Kühlbrandt. 2014. The Resolution Revolution. *Science*: vol. 343, no. 6178, pp. 1443–1444, 28 March.
84. W. Kühlbrandt. 2014. Cryo-EM enters a new era. *eLife* 1 January: e03678.
85. N. Guerrini. 2012. Design and characterization of a highly miniaturised radiation monitor. *Proceedings of International Workshop on Semiconductor Pixel Detectors for Particles and Imaging*, Inawashiro, Japan, September 3–7.
86. B. Krieger et al. 2011. Fast, radiation hard, direct detection CMOS imagers for high resolution transmission electron microscopy. *IEEE Nuclear Science Symposium Conference Record*, pp. 1946–1949, Valencia, Spain, October 23–29.
87. Available at http://www.fei.com/accessories/falcon-II/.
88. Available at http://www.directelectron.com/products/de-series.
89. Available at http://www.gatan.com/products/tem-imaging-spectroscopy/k2-direct -detection-cameras.
90. N. Guerrini et al. 2013. Design and characterization of a highly miniaturised radiation monitor HMRM. *Nuclear Instruments and Methods in Physics Research A*, vol. 731, pp. 154–159.
91. S. Gunes-Lasnet. 2013. HMRM: A highly miniaturised radiation monitor as engineering tool supporting satellite design. *Proceedings of Radiation and its Effects on Components and Systems Conference*, Oxford, United Kingdom, September.
92. E. Mitchell et al. 2014. The highly miniaturised radiation monitor. *JINST*, vol. 9, P07010 July.
93. Available at http://www.sstl.co.uk/Missions/TechDemoSat-1—Launched-2014/Tech DemoSat-1/TechDemoSat-1—Payloads.

11 Technology Needs for Modular Pixel Detectors

Paul Seller

CONTENTS

11.1 Introduction 265
11.2 Hybrid Semiconductor X-Ray Imaging Detector Materials and Contacts...266
11.3 ASIC Technology for Hybrid Pixel Detector Readout 269
11.4 Area Interconnect 270
11.5 Noise...272
11.6 Cooling 274
11.7 Dead Areas in Modular Systems...274
11.8 Through-Silicon Via for I/O Connection Needed for Low Dead Areas 276
11.9 3DIC Technology Advantages 277
11.10 Flatness and Stress...279
References...279

11.1 INTRODUCTION

Area-array imaging detectors are ubiquitous in our everyday lives and in science instrumentation. For imaging infrared, visible, and ultraviolet photons, the usual approach taken is to use focusing optics to concentrate the light onto a small imaging area and to use the smallest pixels possible to reduce the cost of the imaging detector. This is usually true in science experiments as well as commercial cameras. The modality for detecting x-rays is very different. In most cases, it is impossible to focus the x-rays, and large-area detectors are required. This is true for x-ray photon and particle detectors in applications from high-energy physics, space science, and synchrotron applications. Many applications can use large, flat panels[1] with deposited scintillators or scintillating screens. Amorphous silicon screens have been very successful in medical and security applications as well as science but do have limitations in speed and several other performance criteria.[2] Complementary metal–oxide semiconductor (CMOS) sensors have recently surpassed charge-coupled devices for many applications and can now be constructed at a wafer scale (see Figure 11.1).[3] Even with these wafer-scale sensors, there is a requirement to tile these to create larger arrays. Up to a certain size, this is relatively easy as the detectors can be three-side butted and connected to scintillator screens. The readout can be performed on one edge using conventional wire bonding, and the sensor area can be connected to a stable substrate that also allows cooling if required.

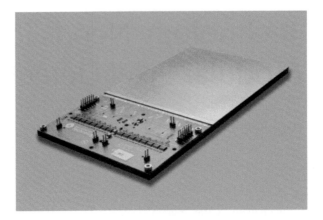

FIGURE 11.1   The three-side, buttable LASSENA CMOS sensor designated at STFC 140 × 140 mm with 50-μm pixels mounted on a readout card.

These CMOS sensors are also used in this configuration for the direct conversion mode of detecting x-rays using diodes that are integrated in the CMOS technology to convert the x-rays to electrons. This is usually restricted to low-energy x-rays (<10 keV) and can suffer from radiation damage effects. Overall, flat panel detectors and CMOS sensors have been very successfully integrated into arrays using large-area modules. The situation with direct-conversion semiconductor detectors using thicker conversion layers is very different. This requires the use of separate conversion layers of thick silicon or high-atomic-number material that is bonded to the readout electronics in so-called hybrid configuration. The process technologies and interconnections for these hybrid sensors are much more complicated.

11.2   HYBRID SEMICONDUCTOR X-RAY IMAGING DETECTOR MATERIALS AND CONTACTS

X-rays interact inelastically with matter in three ways: (1) photoelectric, (2) Compton, and (3) pair production.[4] These energy loss mechanisms all eventually produce thermalized electron–hole pairs in the conduction and valence bands of the semiconductors. The basic conversion process is that the electric field of the x-ray photon transfers energy to a bound electron in the material. This can, for instance, be a k-shell electron, which causes it to be ejected out of the atom. This resulting photoelectron (and subsequent Auger electron) interacts with the electric field in the crystal lattice–elevating electrons to the conduction band leaving holes in the valence band of the semiconductor. This is a statistical effect producing electron–hole pairs by multiple photon–phonon interactions. On average, one electron–hole pair is produced by 3.6 eV of the photoelectron energy in silicon. These mobile carriers are drifted in an applied electric field across the detector. In semiconductor detectors, a large voltage can be applied across the crystal to drift these charge carriers to the anode and cathode sides of the detector. As the charge carriers move, they induce a charge on the electrodes on the anode and cathode of the detector. It is this induced signal that is measured by the preamplifiers on the application-specific integrated circuits (ASICs), as explained in Section 11.3. Detectors can have thousands or millions of contacts and similar numbers of amplifier channels. In most cases, the amplifiers are fabricated on silicon ASICs, with each amplifier connected directly to the detector contact, as explained in Section 11.4.

Silicon has been used as good conversion material to convert x-rays into charge carriers. It has a large bandgap, so very few electron–hole pairs are thermally produced at room temperature. Blocking contacts can be produced, which further reduce unwanted leakage currents. The bandgap is acceptably low so that the photoelectron produces many carriers. Good crystals with low trapping/recombination centers are produced so that the charge carriers produced can drift large distances in the material. The advanced technology allows nearly any geometry of pixels and contact structures. The photoelectric effect dominates in most materials up to a few tens of kiloelectronvolts. High-resistivity silicon detectors are typically limited to approximately 500-mm thick material (at reasonable full-depletion voltages), and this thickness can absorb most photons up to approximately 15 keV by the photoelectric effect. Unfortunately, above this energy, silicon becomes essentially transparent. Photoelectric absorption increases as the fourth power of the atomic number so that higher atomic number elements can absorb x-rays more efficiently. What we would like is a high-Z material (for good absorption) with a high bandgap (for low leakage) and good crystal quality with good charge transport (high mobility–lifetime product). As one would expect, the perfect high-Z material does not exist. Germanium is nearly a perfect material with a reasonable atomic number and superb charge transport, but it has a low bandgap of 0.66 eV, which requires cooling to cryogenic temperatures to stop unacceptable leakage current. GaAs and HgI have been proposed for a long time,[5] but the material production has not produced reliable material with unpopulated bandgaps. Another continuing problem with epitaxial GaAs is that detector-grade material is needed to be grown on thick substrates, which then have to be removed. This process has been a technical barrier to the development of material with good conversion efficiency. Currently, highly doped bulk material has acceptable performance in some applications.[6] CdTe and CdZnTe are currently the most favored crystals at this moment in time for this application. The bandgap is approximately 1.4 eV, increasing with increasing zinc concentration. The advantage of a wider bandgap is to produce higher resistivity and lower leakage currents. As a rule of thumb, the energy to produce an electron–hole pair in semiconductor material is approximately three times the bandgap, and this holds true for Cd(Zn)Te. The charge transport over a few millimeters is acceptable, and growth techniques have now improved to give reliable material.* The x-ray absorption[7] is shown in Figure 11.2 so that several millimeters of material can efficiently convert 100-keV photons and above.

Area array detectors built with CdTe or CdZnTe (CZT)[8,9] typically use 1–10-mm thick material depending on the energy of the incident photons and the efficiency that is required. Cd(Zn)Te detector material is now grown with good quality. Bulk material in boules of 4-in. diameter can be obtained by several Bridgman techniques,

---

* Redlen Technologies, KromeK Ltd, Acrorad Co. Ltd.

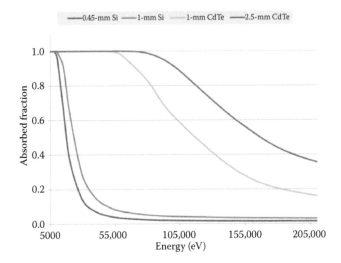

FIGURE 11.2   X-ray absorption against photon energy for different thicknesses of CdTe and Si.

the traveling heater method, or multitube physical vapor transport. Resistivities of 10^{10}–10^{11} ohm-cm and mobility–lifetime products of 10^{-2} to 3×10^{-2} sec are achieved.[10] There is now a good availability of high-quality detector material, but there are still many academically interesting and technically challenging issues to be solved in creating reliable electrical contacts to the material.

Indium and aluminum contacts (cathode and anode, respectively) are used for good-quality Schottky blocking contacts on CdTe.* The advantage of these is that there is reduced leakage current, but by blocking the thermally generated and signal-generated carriers, there are issues due to so-called polarization or field collapse below one of the contacts in the bulk of the material.[11] This can be controlled by pulsing the bias supplies if the problem is not too pronounced. Gold and platinum are used as more ohmic contacts to both CZT and CdTe material. There are still problems of uniformity, injection of current, and stability of contacts. At present, single-element or large pad (1–2-mm) detectors are available with good uniformity from material suppliers and other fabricators.† Small-pixel detectors with lithographic features below 250 μm are very much more difficult to obtain.[12] Detectors with small pixels (50–250-μm pitch) are required not only due to their imaging resolution but also because the small-pixel effect allows single carrier-type readout due to the high weighting field around the pixels. This gives very good spectroscopic capability and reduces the effects of the poor hole mobility in Cd(Zn)Te. In CZT and CdTe, these small pixels are hard to obtain commercially. GaAs detectors are produced with small pixels due to the more advanced processing technology. This modality is one area where GaAs detectors have an advantage over other high-Z detector materials. Germanium historically has not been available with pixelated contacts, but in the recent few years, there are developments of strip and

---

* Acrorad Co Inc.
† Creative Electron Inc. San Marcos, CA, United States.

pixel-contacted detectors,[13] but still there is the huge issue of running the detector at liquid nitrogen temperature and readout electronics at a higher temperature.

11.3 ASIC TECHNOLOGY FOR HYBRID PIXEL DETECTOR READOUT

Clearly, for large-area imaging detectors with many imaging pixels, there is a need to have a high level of segmented multichannel readout. Several decades ago, the only way to achieve this was via massive fanout schemes to route signals to discrete low-density electronics. At present, CMOS technology allows us to build very dense, low-power electronics with many channels, which can be bonded directly to the conversion/sensing medium.[14] There are different requirements for the CMOS technology that is used for the analog front-end signal processing, as opposed to that for the digital signal processing and off-chip communication. For the analog part of the electronics, there is a requirement for a robust technology that has low electronic noise in terms of thermal noise, junction leakage and capacitance, and flicker noise. Also, there is a need for large power supply voltages (3.3–5 V) so that it can accommodate a wide dynamic range. For digital signal processing and communication, there is a requirement for very-high-speed, high-density, low-power technologies, which are compatible with the more modern low-voltage-supply, deep-submicron processes.

There seems to be a technology optimum at present at around 0.35–0.25-μm minimum feature size for the analog requirements. These CMOS processes have robust insulated-gate field-effect transistor (FET) oxides so that input protection circuits can be optimized for noise performance and that they have wide dynamic range and low-noise coefficients. The additional advantage of these technologies is that they are relatively cheap costing less than 100,000 euro for an engineering run of a new design and 2000 euro per wafer. This technology is also relatively easy to design with and has acceptable die sizes of approximately 20 × 20 mm. Obviously, the large feature size limits the complexity of circuitry that can be integrated in a pixel, but the device performance compensates for this. Surprisingly, even with the relatively thick gate oxides, the total dose radiation hardness of these technologies can be acceptable,[15] and with suitable consideration of substrate biasing, latch-up-free designs can be achieved.

In comparison, the digital signal processing side of the measurement chain can benefit from the rapid development of deep-submicron processes. Science applications have used 0.25, 0.18, and 0.135 μm and are now using 65-nm[16] and proposing 22-nm technologies for digital components. These technologies are well suited to high-speed analog-to-digital converter (ADC) architectures and to very fast data manipulation for data specification and packaging. The technologies have their own limitations in terms of gate oxide thickness, noise, and cost. The nanometer gate oxides and small interdevice distances reduce the power supply range and can introduce noise problems. A major practical problem is the cost of a mask set and engineering wafer batch and the design complexity, which extends to design engineer costs. A full ASIC development can easily cost over a million euro and extend over two years, which can be a significant problem for the budget of some scientific experiments.

These high-density processes will be important for new devices, and SiGe and multigate FET processes will expand the range of processing architectures that we can integrate close to the front end of detector systems. Also, new auto-gain switching front-end circuits[17] allow wide dynamic range without needing large power supply voltages. In Section 11.9, we will discuss how we can use mixed technologies to optimize the use of the strong points of these different technologies.

11.4   AREA INTERCONNECT

Pixel detectors require a direct connection from the pads on the sensor material to the bond pads on the ASICs. Figure 11.3a shows a simple pad geometry for connection to the detector where the pixel pitch on the readout ASIC is the same as the detector pixel pitch. It is also possible to fan out the connections on the detector with multilevel metal routing on the detector, as shown in Figure 11.3b. This has to be done very carefully as there is a huge scope for signal crosstalk. With integrating readout and synchronous input signals,* where the signal is totally removed from the detector, this might not be a problem, but transients can still upset thresholds in these systems.

The pitch of x-ray imaging systems currently ranges from approximately 50 µm to 1 mm. Bump bonding is used to connect the detector pixels to the ASICs. There are many different technologies to do this depending on the requirements of the detectors and environmental constraints.

The industry-standard area bump-bonding method is to deposit solder onto underbump metallization on the pads of the detector and ASIC and then to align the two and heat them to reflow the solder. Various solders are used including lead–tin, bismuth–tin, indium, and silver alloys depending on the temperature to reflow and the operating temperature that is required.[18] Typically, these materials require 240°C–140°C to reflow. Indium is used either in a lower temperature reflow process or straight compression bonding[†] and gives good results but cannot be used if high operating temperatures are ever experienced. With all these processes, either fluxes or special gas environments including nitrogen, hydrogen, or formic acid have to be used to ensure good contact between parts. These processes can allow 10-µm bond diameters to be used, which allows 25–50-µm pixels to be connected.[19] Another method used particularly for larger-pixel (150–250 µm) CZT detectors is gold-studded and silver-loaded epoxy dots. The gold studs are either bonded or deposited on the ASIC part, and the silver-loaded epoxy is screen-printed on the detector. A process for this is used by our group at the Science and Technology Faculty Council (STFC) for the connection of CZT. After alignment, the unit is heated to set the epoxy. This can be done at low temperatures of 80°C–120°C so that the detectors are not affected. CZT is a very sensitive crystal, and it has been seen that temperatures of around 150°C can cause movement of impurities and damage sites, which affect detector performance. This method uses 50-µm gold studs, which give large distances between the components and are good for capacitance and stress relief. For all these processes, there is the option to use an underfill after bonding to strengthen the connection and seal against corrosion.

---

* European XFEL, Hamburg. Available at http://www.xfel.eu/.
† VTT. Technical Research Centre of Finland. Available at http://www.vttresearch.com.

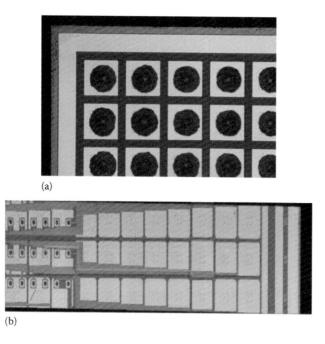

FIGURE 11.3 (a) Hexitec CdTe 250-µm pitch pad geometry with silver glue dots. (b) European X-ray free electron laser LPD redistributed bonding pads 250 × 400 pitch (left) to the 500-µm pitch pixels on a two-layer interconnect silicon detector.

Other area interconnect methods have been used such as direct gold-to-gold and indium pillar contacts. Also, solid–liquid interface diffusion (SLID) bonding can be used. This Cu-Sn process leaves a bond that is robust to subsequent processing up to 400°C. This is being used in 3D-integrated circuit (3DIC)[20] and silicon interposer technology but could be a prospect for a high-strength detector to ASIC connection.

A serious issue with hybrid detector systems is stress that is produced in the detector material and to the interconnection surfaces.[21] One inherent issue is the different coefficient of thermal expansion (CTE) of the different components. The bump-bonding interconnection processes above require the module to be heated to 100°C–200°C and above. The device might then be required to operate at a 0°C–10°C temperature for best performance (with CdTe). With large detectors of 20-mm length and above, this causes a considerable strain to be set into the interconnection face. With the CZT detector connected to silicon, this is several microns of linear change over a few centimeter detector with 100°C temperature change. Either the bumps have to accommodate this or the material is stressed, and the structure could bend like a bimetallic strip or break. None of these are good, but considering using the minimum processing temperature and allowing the bumps to take, deformation is the least bad option. With CZT, there is certainly an additional problem with inbuilt stress as this can cause residual internal electric fields around the pixels as well as the long-term reliability of the fragile electrical contact interfaces.

Another problem is the connection of the ASIC to the cooling/alignment block that is used to support the ASIC to the instrument. Aluminum would be a very convenient mounting substrate but has a very big difference in CTE from silicon. Molybdenum is often used as it has a good CTE match to silicon (even better to CZT) and is a good thermal conductor. It is, however, quite expensive. A good option we have found is to accept the mismatch with a metallic mounting plate but using a compliant heat transfer tape. We had a spectacular failure of a 54 × 54-mm area CMOS detector when hard-glued to a copper block but a complete success when mounted with compliant tape. With smaller ASICs, the epoxy glue is quite adequate but can still stress the silicon and effect reliability.

11.5   NOISE

There are several intrinsic noise processes that limit the performance of pixelated detectors. These can be split into sources that are related to the detector and processes that are related to the way the electronics reads out the detector.[22] These processes define the ultimate limit of performance, but there are many ways that the architecture can be designed inappropriately so that these are not the dominant effect.

Assuming that all environmental noise is eliminated, the fundamental front-end related noise can be summarized as equivalent noise charge (ENC), which is given by the formula

$$\mathrm{ENC}^2 = \mathrm{ENC}^2_{\mathrm{thermal}} + \mathrm{ENC}^2_{\mathrm{shot}} + \mathrm{ENC}^2_{\mathrm{flicker}}$$

$$\mathrm{ENC}^2 = \left(b_t c_{\mathrm{tot}}^2 kT \frac{2}{3gm} + b_s iq + b_f c_{\mathrm{tot}}^2 \frac{Q_f}{gm^2} \right) [\mathrm{coulombs}^2]$$

The definition of terms are fully explained in the reference: (b_t, b_s, b_f) are coefficients that are calculated for a particular shaping filter, c_{tot} is the capacitance of the detector and tracking, i is the leakage current of the detector, and gm is the transconductance of the input FET of the amplifier. T and k are the temperature and the Boltzmann constant.

The ENC is the signal charge that you would have to inject onto the detector in order to produce an output signal that is equivalent to the rms noise that you measure at the output of the system. The terms for the individual ENC components are added in quadrature to give the total noise. This method of addition is required as statistically independent noise powers (voltage^2) add. These noise contributions from the electronics affect the accuracy with which the signal can be measured. The signal produced in the detector has an intrinsic statistical noise. In semiconductor detectors, the ionizing radiation produces a number of charge carriers, as described in Section 11.2. With a stochastic process, one would expect the noise on the number of carriers to be the square root of the number of carriers. Actually, because of the carrier formation process, the signal noise is better than this square route by a factor that is known as the Fano factor.[23] This term is the ultimate or Fano limit of resolution of the detector but has to be added in quadrature to the electronic noise contributions. Lowering the bandgap increases the number of carriers, which improves the accuracy of measurement, which is one reason that Ge detectors have good resolution. The downside of low bandgap is that the thermally generated leakage current can offset this advantage. High-Z conversion materials all have other noise mechanisms and inhomogeneity, which produce signal uncertainty problems. In CZT, the major problem areas can be small inclusions,[24] other bulk and contact nonuniformities that cause very localized changes in charge carrier transport efficiency. This creates a fixed-pattern noise, which unfortunately cannot always be removed in subsequent processing. Other problems occur due to charge sharing across pixels.[25] This can cause additional signal uncertainty problems that are related to the exact detailed system parameters and subsequent signal processing.

If one inspects the basic ENC formula, one can obtain several design objectives. As the thermal noise (b_t) decreases with shaping time and leakage current (shot) noise (b_s) increases with shaping time,[26] there is an optimum shaping time for a detector:

$$\mathrm{ENC}^2 = \left(\frac{e^2}{2\tau} c_{tot}^2 kT \frac{2}{3gm} + \frac{e^2\tau}{4} iq + \frac{e^2}{2} c_{tot}^2 \frac{Q_f}{gm^2} \right) [\text{coulombs}^2] \, (\text{for CR}-\text{RC})$$

Because of the formula for the thermal noise of an insulated gate field effect transistor (IGFET),

$$\mathrm{ENC} = (c_d + c_{in}) \sqrt{b_t kT \frac{2L}{3\sqrt{2\mu I_D C_{in}}}} \, [\text{coulombs}]$$

One can optimize against c_{in}, and obtain an optimum input FET gate capacitance of between one-third and one time the total input detector capacitance.[27] This so-called matching condition sets a good criterion for the input circuit design.

Further inspection of the ENC formula shows an even more important design driver. If the leakage current is reduced, the leakage noise reduces. If the detector capacitance is reduced, the thermal noise reduces. The way to achieve both of these together is to segment the detector into smaller pixels. A limit is reached where capacitance actually increases, but to the first order, having many pixels improves measurement accuracy. Also, as there are more amplifier channels, the possible bandwidth to measure signal photons is increased. Another practical limitation is reached when the pixel size is so small that charge carrier clouds spread across multiple pixels, and this charge-sharing effect then has to be resolved by more complicated signal processing. Overall, fine pixel systems have the potential to have lower noise and be faster than single-channel systems.

Several direct-conversion x-ray imaging systems have been built using a photon-counting technique, including Medipix, Pilatus, and XPAD.[28] Typically, this involves the front-end electronics amplifying the charge signal and then using a discriminator to count the number of hits in the pixel with a value above a set level. Some ASICs have the capability to have several discriminators and several counters to get a rudimentary spectroscopy.[29] To achieve a true spectrum, the threshold has to be scanned during repeated experiments. Another method used by STFC is to amplify the signal and output the analog signal for every photon incident in the pixel and perform an ADC conversion on this value.[30] The advantage of this is that the position and energy of every charge cloud are measured allowing an effective improvement in efficiency and good correction of charge sharing. The advantage of this hyperspectral data with very good spectral resolution is that there is a new richness in the data to allow advanced imaging techniques.[31]

11.6 COOLING

An important technology aspect of array detectors is the close integration of the electronics to the semiconductor detector material. As we have seen, this is important for capacitance and other noise injection possibilities. With wire-bonded detectors, it is possible to have thermal breaks between the electronics and the detectors.[32]

The intimate connection of the electronics to the detectors in array detectors means that the electronics and the detectors are essentially thermally coupled. The low-noise front ends and complex electronic processing generate a lot of heat; 0.5 w/cm^2 is certainly possible. CZT semiconductor detectors will work at room temperature, but the leakage increases rapidly with elevated temperature. In order to operate the detectors efficiently, some form of heat removal is required. In most systems, this is only achieved by attaching the ASICs to a cooling substrate. The mechanics and materials used for this are distinct in different systems ranging from circulating binary ice, high-pressure evaporative liquids,[33] Peltier devices, forced air-cooling or heat pipes, to off-detector cooling. The technologies vary, but the consequence is the same in that the back of the detector becomes very congested. The cooling takes place in exactly the space that one would like to route interconnections to control and take data from the ASICs. An example of the cooling is shown in Figure 11.4, but many other configurations are used.[34]

11.7 DEAD AREAS IN MODULAR SYSTEMS

As already stated, large-area hybrid detector systems have to be built from multiple modules. There is a general requirement to reduce the dead areas between modules

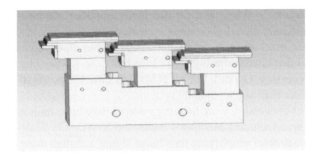

FIGURE 11.4 Cooling block of the multimodule Hexitec system.

for cosmetic reasons in the image as well as necessary science issues for full image coverage. Several hybrid systems are built using conversion material that is directly bump-bonded to ASICs. In order to read out these ASICs, at least one edge of the ASIC has to be exposed (not covered by conversion material) so that wire bonds can be made to pads on the surface. An image of this is shown in Figure 11.5.

A gap has to be left between detectors to allow room for these wire bonds and the connection board; typically, this is approximately a minimum of 1–2 mm. This has two consequences in that there is a dead area for the detector, and the ASIC is directly exposed to radiation, which can have radiation damage consequences. One possibility is to have a roof-tile geometry for the detectors that is shown in Figure 11.4. This allows the wire-bonded area of one ASIC to be covered by an active detector area of the next detector. This is very effective but produces a detector surface, which is not a flat surface normal to the incident x-rays. Section 11.8 describes a new method to allow full coverage of the detector plane and to have a flat detector. All direct conversion detectors need a high-voltage bias to be connected to the back of the detector. This can also be done in the same method as the wire bond to the ASIC input/output (I/O) pads, as shown in Figure 11.6.

The other three edges of the ASIC, not required for bonding, can be covered by the detector material so that the detectors can be spaced as close as mechanically allowed. The edges of the detector also causes dead areas. Semiconductor detectors have to be cut from a larger wafer, which nearly always causes the edge material to have significantly different properties to the bulk material. In silicon, a diced edge has crystal damage, which shorts the top and bottom of the edge together. So-called edgeless or active-edge silicon detectors have been built by very specifically fabricating the edges.[35] With GaAs and CdTe detectors, usually, there is a guard band on the edges of the detector to stop the edge leakage current from entering the edge pixels. STFC has been working on active-edge CdTe detectors, which have shown

FIGURE 11.5 Wire bonds connecting the ASIC to the readout board. The detector (top) cannot be located close to these wire bonds, so there is an area of the ASIC that is not covered by the detector. This causes a large dead area next to the detector.

FIGURE 11.6 Wire bonds to the exposed I/O pads on the fourth side of the ASIC, with a single high-voltage connection to the back of the detector.

that good counting efficiency and spectroscopy can be achieved without guard bands allowing very close packing of detectors.[36] Similar results are shown on Medipix detectors with very small intermodule dead regions.[37]

11.8   THROUGH-SILICON VIA FOR I/O CONNECTION NEEDED FOR LOW DEAD AREAS

STFC has built an x-ray imaging spectroscopy readout system based on CdTe detector material. A 20 × 20 × 1-mm crystal is bonded to a Hexitec ASIC with gold stud bonds to 80 × 80 pixels of the detector. Each pixel is 250 × 250 µm and has an amplifier and a peak track-and-hold circuit. This allows the energy of every photon to be recorded and measured with sub 1-keV full width half maximum accuracy from 4 to 200 keV and above.[38] The first versions of the ASIC allow the detector modules to be butted on three edges with only a 170-µm gap between the detectors. The fourth edge of the ASIC has wire bonds, which precludes butting the detectors in a flat geometry. A similar restriction has compromised the design of the popular DECTRIS PILATUS detector. The only way to avoid these wire bonds restricting the coverage of the ASIC is to read out the ASIC from the back of the detector. Several methods are proposed, but STFC has chosen an approach where the wire bond pads are shielded with a metal plane on the top of the ASIC to stop electromagnetic injection into the detector. The I/O bond pads are then read out with through-silicon vias (TSVs).[39] The process involves thinning the wafer to 100 µm so that it is still mechanically robust and then etching a TSV from the back to connect to the back of the metal bond pad. The TSV is plated with metal, and a pad of the same metal is fabricated on the back of the ASIC. Then, an almost identical but mirrored wire bonding method is used to connect to the back of the *4S* ASIC, as shown in Figure 11.7.

With this geometry, it is possible to extend the active CdTe over the whole of the front surface of the ASIC. This allows large area coverage with four-side butting and only very small dead regions on all the edges of the detector modules.

FIGURE 11.7 Wire bonding to redistributed pads on the back of a 4S ASIC. Each pad has a TSV through the thinned 100-μm ASIC.

11.9 3DIC TECHNOLOGY ADVANTAGES

CMOS ASIC technology has progressed over the decades with the relentless Moore's law. However, as we approach the 22-nm feature-size regime, it is believed that simple dimension scaling will not be sufficient. This is where the so-called 3DIC technology will be an important next-generation technology.[40] The objective of 3DIC technology is to distribute the signal processing on several layers of silicon and then connect the layers by TSV technology, as shown in Figure 11.8. The advantage comes partly in simply having multiple layers of silicon but mostly due to the immense connection density. This mimics the biological anatomy of the animal nervous system with the immense processing capability that is produced by the billions of 3D neural connections. The drive for this technology is to scale processing power, but this will also be extremely useful for area array detectors. As we have seen, different CMOS processes have advantages for either analog or digital functions. Connecting them together allows the best technology to be used for the analog process and optimized for the digital process. For instance, analog processes need wide dynamic range

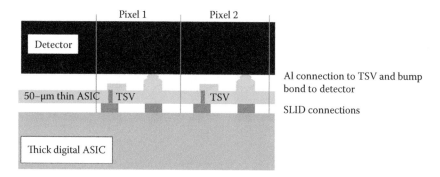

FIGURE 11.8 3DIC stacking showing bump bond from detector to active surface of the thin ASIC. The TSV can be seen connecting to the SLID bond to the active layer of the thick ASIC. In this case, there is a SLID support under the bump bond.

and low noise whereas digital technologies need small feature size and high speed. Typically, connections would be made between the layers within each pixel. The signal from the analog layer is amplified and transferred using a small via from this top layer through the thinned (20–50-μm) silicon substrate to the digital layer below.

There are several demonstration prototypes of this technology being trialed in the science community.[41,42] The 3DIC technology is only starting to be used in the commercial sector,[43] and, really, the first large investments in plant are only now being made to have capability in the next few years. STFC has a program to investigate the feasibility of 3DIC technology for imaging detectors.[44] We have used the analog front-end part of the Hexitec CZT readout ASIC as the pixel for the analog layer. The amplified output from each 250-μm pixel is then sent via a TSV to a peak track-and-hold and Wilkinson ADC circuit on each pixel of a lower digital ASIC. The aim of this is to digitize all the signals from the detector on one 3DIC ASIC. The analog and digital ASICs are both fabricated on an Austria Mikro Systeme 0.35-μm CMOS process to reduce the cost of the trial. The tungsten via technology and the SLID interconnection between the wafers are performed by the EMFT Fraunhofer institution in Munich.[45] This technology has rather large TSVs of 3 × 10 μm and requires a large connection pitch due to the large size of the SLID pad technology that is shown in Figure 11.9. The advantage of the process is that it is a via-last (or back-end-of-line) process so that any CMOS process can be used, and there are possibly different technologies for the two layers. TSV processing can induce a large strain in the silicon, which can affect the active circuitry, particularly with modern processes that rely on strain in the FET for high-performance operation. The relatively large space around the via in this technology is an advantage in this respect.

The aim of 3DIC technology for imaging detectors would be to eventually reduce the vias to a few microns and have pixel sizes in the range of 50–100 μm. This would then allow a fine pitch detector with smart signal processing in each multilayer pixel. In order to do this, we will probably have to wait for extremely thin silicon layers and

(a)   (b)

FIGURE 11.9 (a) The four 3 × 10-μm TSVs are shown in the area of the ASIC containing the 3DIC TSVs. The aluminum (AL) contact to this can be seen entering from the top. The other large square is the bump bond pad for connection to the detector. (b) The SLID bond pairs connecting to the digital ASIC: one of the pairs is the TSV connection, and one is a pad under the bump bond pad for support.

the introduction of front-end-of-line processes that have the vias inserted in the silicon before active devices are fabricated. These processes will be very expensive and probably be driven by the digital processing and communications markets. This will mean that the technologies might not be ideal for all science applications. Until then, via-last and silicon-interposer* technologies will be used, which will allow mixed technologies and cheaper cost if not ideal in other respects. The alternative silicon interposer technology has been used in the large pixel array detector (LPAD) detector for the XFEL European x-ray source.[46] This is a useful technology, but is widely recognized as a stepping-stone technology to full 3DIC technology for these imaging applications.

11.10   FLATNESS AND STRESS

Silicon wafers have to be very flat in order for the photolithographic and other processes to be reliably performed on the active surface. This dictates the use of very thick wafers of 700–1000 μm for dimensional stability. These thick wafers are perfect for subsequent bump-bonding processes to detectors, which require a relatively relaxed flatness of a few microns across the ASIC. In conventional solder bump bonding of silicon to detectors, the bonding can be done at the wafer level or at the ASIC level. When the wafer is thinned, the highly stressed processed side of the ASIC causes the whole wafer to distort. In 3DIC processing, the only way to obtain very thin and flat wafers is to attach them to handle wafers and perform all the grinding and interconnect steps, while the thin wafer remains on the handle wafer.

If we want to use these devices for imaging detectors, the final stack has to have at least one thick substrate to remain flat enough for the subsequent bonding to the detectors. The stress in the stack still remains, but, at least, the ASICs will be flat enough to be bonded.

An even more difficult challenge is when TSVs are used on a single-layer device to redistribute the I/O connection from the ASIC bond pads to the back of the wafer, as described in Section 11.8. This is particularly useful for four-side bonded area array detectors with very small dead areas. At STFC, we have used single layers that were thinned to 100 μm and seen that the stress in the ASIC causes them to distort over an area of 20 mm. The solution to this might be to firmly bond them to the detector. The detector and the bonding will then form the support for the structure.

REFERENCES

1. S. Kasap. 2011. Amorphous and polycrystalline photoconductors for direct conversion flat panel x-ray imaging sensors. *Sensors*, 11, 5112–5157, ISSN 1424-8220. Available at http://mdpi.com/journals/sensors.
2. J.A. Seibert. 2006. Flat panel detectors: How much better are they? *Pediatr. Radiol.* 36(Suppl 2), 173–181, September.
3. I. Sedgwick. 2013. LASSENA: A 6.7 megapixel, 3 sides buttable wafer-scale CMOS sensor using a novel grid-addressing architecture. *Proceedings of the 2013 International Image Sensor Workshop*, June 12–16, Snowbird, UT, United States.

---

* Silex Microsystems AB. Available at http://www.silesmicrosystems.com.

4. G.F. Knoll. *Radiation Detection and Measurement*, Chapter 2. ISBNO-471-49545-x, John Wiley & Sons.
5. B.E. Patt. 1995. Development of mercuric iodide detector array for medical imaging applications. *NIM A* 366, 173–182.
6. O.P. Tolbanov. 2001. GaAs structures for x-ray imaging detectors. *NIM A* 466, 25–32.
7. X-ray interactions with matter. Available at http://henke.lbl.gov/optical_constants/.
8. P. Seller et al. 2011. Pixellated Cd(Zn)Te high energy x-ray instrument. *JINST* 6, C12009, December.
9. W. Kaye et al. 2010. Calibration and operation of the polaris 18-detector CdZnTe array. *Nuclear Science Symposium Conference Record (NSS/MIC)*, 2010 IEEE.
10. M.C. Veale. Cadmium Telluride and Cadmium Zinc Telluride Based Detector Systems.
11. A. Cola. 2009. The polarization mechanism in CdTe Schottky detectors. *Appl. Phys. Lett.* 94, 102113.
12. M.C. Veale, S.J. Bell, P. Seller, M.D. Wilson and V. Kachkanov. 2012. X-ray microbeam characterization of a small pixel spectroscopic CdTe detector. *IOP JINST* 7, P07017.
13. T. Krings et al. High resolution x-ray spectroscopy. *JINST* 9, May 2014.
14. R. Ballabriga. 2011. Medipix3: A 64 k pixel detector readout chip working in single photon counting mode with improved spectrometric performance. *NIM A*, 633 (Suppl 1).
15. F. Faccio. 1998. Total dose and single event effects (SEE) in a 0.25 m CMOS technology. *4th Workshop on Electronics for the LHC*, Rome, CERN/LHCC/98-36, October.
16. S. Bonacini, P. Valerio, R. Avramidou, R. Ballabriga, F. Faccio, K. Kloukinas and A. Marchioro. 2012. Characterization of a commercial 65 nm CMOS technology for SLHC applications. *JINST* 7, P01015.
17. J. Becker. 2013. Challenges for the Adaptive Gain Integrating Pixel Detector (AGIPD) design due to the high intensity photon radiation environment at the European XFEL. PoS Vertex 2012 012.
18. C. Lee et al. 2008. Development of low temperature bonding using In-based solders. *Electronic Components and Technology Conference 2008 ECTC 2008* 58th, 1295–1299.
19. P. Norton. 2002. HgCdTe infrared detectors. *Opto-Electron. Rev.* 10 (3), 159–174.
20. P. Ramm. 2008. 3D-IC fabrication challenges for more than Moore applications. *Workshop Manufacturing and Reliability Challenges for 3DIC Using TSVs, EMFT*, pp. 841–846, San Diego, September 26.
21. W. Sang. 2004. Primary study on the contact degradation mechanism of CdZnTe detectors. *NIM A* 527, 487–492.
22. P. Seller. 1996. Noise analysis in linear electronic circuits. *NIM A*, 376, 229–241.
23. A. Owens. 2002. On the experimental determination of the Fano factor in Si at soft x-ray wavelengths. *NIM A* 491, 437–443.
24. A.E. Bolotinkov. Internal electric-field-lines distribution in CdZnTe Detectors measured using x-ray mapping. *IEEE Trans. Nucl. Sci.* 56 (3), 791–794.
25. E.N. Gimenez. 2011. Study of charge sharing in Medipix3 using a micro-focused synchrotron beam. *JINST* 6 C01031, *12th International Workshop on Radiation Imaging Detectors*.
26. P. Seller. 1999. Summary of thermal, shot and flicker noise in detectors and readout circuits. *NIM A*, 426 (2–3), 538–543.
27. P. Seller. The matching condition for optimum thermal noise performance of FET charge amplifiers. RAL-87-063.
28. P. Pangaud. 2008. XPAD3-S: A fast hybrid pixel readout chip for x-ray synchrotron facilities. *NIM A* 591 (1), 159–162.
29. C. Ponchut. 2011. MAXIPIX, a fast readout photon-counting x-ray area detector for synchrotron applications. *JINST* 6 C01069, January.

30. L. Jones, P. Seller, M. Wilson and A. Hardie. 2009. HEXITEC ASIC—A pixelated readout chip for CZT detectors. *Nucl. Instrum. Methods Phys. Res. A* 604, 34, June.
31. P. Seller et al. 2011. Pixellated Cd(Zn)Te high energy x-ray instrument. *JINST* 6, C12009, December.
32. J. Headspith. 2007. First experimental data from XH, a fine pitch germanium microstrip detector for energy dispersive EXAFS (EDE). *Nuclear Science Symposium Conference Record*, NSS IEEE, pp. 2421–2428, Hawaii, October.
33. A. Nomerotski. 2013. Evaporative CO2 cooling using microchannels etched in silicon for the future LHCb vertex detector. Available at http://arXiv.org>physics>arXiv:1211.1176, February.
34. C. Broennimann. 2005. The Pilatus 1M detector. *J. Synchrotron Radiation*, ISSN 0909-0495, November.
35. G.F. Dalla Betta. 2012. Recent developments and future perspectives in 3D silicon radiation sensors. *JINST* 7 C10006, pp. 120–130.
36. D.D. Duarte et al. 2013. Edge effects in a small pixel CdTe for x-ray imaging. *JINST* 8 P10018.
37. T. Koenig. 2012. Imaging properties of small pixel spectroscopic x-ray detectors based on cadmium telluride sensors. *Phys. Med. Biol.* 57, 6743–6759.
38. J.W. Scuffham, M.D. Wilson, P. Seller, M.C. Veale, P.J. Sellin, S.D.M. Jacques and R.J. Cernik. 2012. A CdTe detector for hyperspectral SPECT imaging. *IOP JINST* 7, P08027.
39. P. Seller, S. Bell, M.D. Wilson and M.C. Veale. 2012. Through silicon via redistribution of I/O pads for 4-side buttable imaging detectors. *IEEE Proceedings Nuclear Science Symposium*, R04-38, pp. 4412–4146, Anaheim California.
40. 2013. 3DIC proceedings. *IEEE*, October 2–4, San Francisco.
41. E. Ramberg. 2013. 3-dimensional ASIC Development at Fermilab. *Vienna Instrument Workshop*, February 14. Available at https://indico.cern.ch/event/186337/session/14.
42. A. Macchiolo. 2012. SLID-ICV Vertical Integration Technology for the ATLAS Pixel Upgrades, February. Available at http://arXiv.org>physics>arXiv:1202.6497.
43. Y. Guillou. 2013. European 3D TSV Summit 2013—Post Show Report. Available at http://www.semi.org/eu/sites/semi.org/files/docs/European%203D%20TSV%20Summit%20 2013%20for%20website.pdf. Last accessed January 6, 2016.
44. Through Silicon Vias for 4-side Buttable Detectors, IEEE NSS MIC Conference Proceedings Seattle 2014. RTSD Conference Record R04-38, pp. 4142–4146.
45. S.K. Lim. 2013. *Design for High Performance, Low Power, and Reliable 3D Integrated Circuits.* Springer, ISBN 978-1-4419-9542-1.
46. M. Hart et al. 2012. Development of the LPD, a high dynamic range pixel detector for the European XFEL. *NSS/MIC IEEE*, pp. 534–537, Anaheim, California, USA.

Index

Page numbers ending in "f" and "t" refer to figures and tables, respectively.

A

Active pixel sensors, 27–29, 46
ADC architectures
  cyclic ADCs, 52, 52f
  ΣΔ ADCs, 53, 53f
  for image sensors, 47–53, 62, 66–67
  pipeline ADCs, 49, 50f
  qualification of, 66–67
  ramp ADCs, 49–51, 51f, 56f
  SAR ADCs, 51–52, 52f
  slope ADCs, 49–51, 51f
ADC comparisons, 62, 63t, 64–67, 65t
ADC efficiency, 62, 64f
ADC requirements, 58–62, 58f, 59f, 60f
ADC resolution, 58–60, 58f, 59f, 60f
ADC topologies
  3D integration, 56–58, 57f, 58f
  column-parallel ADCs, 54–55, 54f, 56f
  future integrations, 56–58, 57f, 58f
  global ADCs, 53–54, 54f, 56f
  for image sensors, 53–58
  pixel-level conversion, 55–56, 56f
Advanced CSP–shaper analog signal processor,
      17–20, 17t
Afterpulsing, 150, 158–159, 159f
All-digital phase-locked loop (ADPLL), 112–113,
      113f
Amplification, 4, 18–20, 18f, 83, 147–150
AMT chips, 108–109, 109f
Analog electronics
  ADC comparisons, 62–66, 63t
  ADC efficiency, 62, 64f
  ADC requirements, 58–62
  ADC resolution, 58–60, 58f
  ADC topologies, 53–58
  design considerations, 62
  for HVCMOS sensors, 27–46
  image sensor architectures, 47–53, 62,
      66–67
  for radiation detection, 47–67
  state-of-the-art design, 62
Analog processing channel, 2–6, 2f
Analog signal processing, 1–24
Analog-to-digital converter (ADCs) guidelines;
      see also Image sensors
  ADC architectures, 49–53, 66–67
  ADC comparisons, 62–67

ADC requirements, 58–62
ADC topologies, 53–58
guidelines for, 47–67
Application-specific integrated circuit (ASIC)
      readouts
  for advanced CSP–shaper processor, 17–22,
      19f–22f
  field-programmable gate array and, 92,
      105–108
  for particle detectors, 1–2
  for pixel detectors, 267–279
  silicon photomultipliers and, 204, 225
Array oscillator, 105–106, 105f
Avalanche photodiode (APD), 142–146, 149f,
      186–187

B

Bias resistance, 35–37, 37f
Breakdown voltage, 150, 154–156, 161, 162f,
      164f

C

Cadmium telluride (CdTe) detectors
  DPP system and, 121, 125–126, 126f
  materials, 142–143
  measurements with, 125, 125f, 132–135,
      132f–137f
Cascode amplifier, 6–10, 7f, 8f, 9f, 84, 84f
Charge-coupled devices (CCDs)
  CMOS image sensors and, 238
  CMOS sensors and, 251, 258
  description of, 2
  noise performance, 186–187
  pixel detectors and, 265
  pixel sensors and, 27
  radiation detection and, 246
Charge-sensitive amplifier (CSA) readout,
      212–217, 212f, 214f, 216f
Charge-sensitive preamplifier (CSP), 2–7, 7f,
      10–12, 17–18, 122–125, 128f, 131,
      229
CMOS active pixel sensors, 27–29, 28f
CMOS sensors
  architecture, 249–256
  conduction band, 241–242, 241f, 245f, 266
  description of, 1

design of, 265–266, 266f
diagram of, 250f
electrons, 240–245, 245f, 247–249, 257
electronvolts, 240–241, 244
energy bands, 241–242, 241f, 245f, 266
energy resolution, 266
growth of, 71–72
HVCMOS process, 257, 257f
INMAPS process, 252, 252f
monolithic CMOS, 238–239
MOS transistor, 244–248, 246f, 248f
NMOS transistor, 244, 244f, 251–252
pixel detectors and, 257–258
pixel sensors and, 27–29, 28f
pixels, 250–253, 251f
p-n junction, 242–245, 242f, 250–251, 257
for radiation detection, 237–258, 239f
readouts, 255–256
row addressing, 250, 253–255, 253f, 254f, 255f
valence band, 241–242, 241f, 245f, 266
CMOS-compatible technologies
comparisons, 193–196
dark count rate, 187, 191f, 195f
dynamic performance, 192–193
isolation, 188–190
noise performance, 187
noise sensitivity, 190–192
photon detection probability, 186, 188, 192–196, 192f, 195f
for radiation detectors, 185–196
SPADs, 186, 188–190, 188f, 189f, 190f, 191f
for time-resolved imaging, 185–187
timing resolution, 193, 194f
trends, 193–196
Coefficient scaling, 82–83, 88
Column-level multiple-sampling ADCs, 73–75, 74f
Column-parallel ADCs, 54–55, 71–72, 72f, 109–110, 111f
Common-source cascode amplifier, 84, 84f
Complementary metal–oxide semiconductor sensors, 1, 27–29, 71–72; see also CMOS sensors
Continuous reset circuits, 39–40, 39f
Correlated double sampling (CDS), 55, 238, 251–252
Crosstalk, 150, 156–157, 157f
Crystal readouts, 141–182, 142f
CSP–shaper design, 6–10, 7f, 17–20, 17t; see also Transconductance
CSP–shaping filter system, 2–6, 2f
CTE detectors, 268, 271–272
Current-mode readout, 222–225, 222f, 223f, 224f
Cyclic ADCs, 52, 52f
CZT detectors, 267–268, 271–274, 278

D

Dark count rate, 150, 152–154, 153f, 154f, 187, 190–194, 190f, 191f, 195f
Delay-line TDC, 99–101, 99f, 100f, 102f
Delay-locked loop (DLL) circuit, 99, 100f, 105–106
Digital pulse-processing (DPP) techniques, 121–138; see also DPP system
Digital signal processor (DSP), 88
Digital-correlated double sampling (DCDS), 72, 74, 85, 238
Digital-to-analog converter (DAC), 33–53, 66, 75, 82–87, 87f, 116f
Displacement effects, 244–245
DPP system
CdTe detectors, 121, 125–126, 125f, 126f, 132–135, 132f–137f
dead time, 129–130
diagram of chain, 125f
DPP firmware, 127
for gamma-ray semiconductor detectors, 122–131, 122f
Ge detectors, 121, 126, 135–138, 137f, 138f
ICR, 123, 124f, 125f, 129–130, 131f, 133f, 134f, 135f, 137f
measurements, 132–138
OCR, 123, 124f, 125f, 129–130, 131f, 135f
performance of, 131–138
PSHA, 126–129
pulse detection, 127, 128f, 130f
working modes, 130–131
for X-ray semiconductor detectors, 122–131, 122f
Dynamic range, 150, 159–160, 160f

E

ΣΔ ADCs, 53, 53f
Electronvolts, 240–241, 244
End-of-column (EoC) cell, 42–46, 46f
Equivalent circuit, 8, 8f, 215, 215f
Equivalent noise charge (ENC), 4–6, 10, 20–21, 225, 272–273

F

Feedback, resistive, 39, 122, 131
Field-effect transistor (FET), 123, 269–273, 278
Field-programmable gate array (FPGA), 88, 92, 105–107, 106f, 126–127
Figure of merit (FoM), 48, 64–67, 65t
Fixed-pattern noise (FPN), 48–54, 60–61, 61f, 66, 74, 88, 238, 273
Flicker noise, 3–4, 249, 269; see also Noise
Floating logic structure, 31

Folded-cascode amplifier, 6–10, 7f, 8f, 9f
Front-end electronics, 203–233; *see also* Silicon
    photomultipliers
Front-end systems, 1–24, 2f

G

Gain, 150, 154–156, 155f
Gain control, 114, 115f
Gain programmability, 21–22, 22f
Gamma-ray semiconductor detectors
    crystal readouts, 175–177
    digital pulse-processing techniques for,
        121–138, 122f
    DPP system for, 122–131
Geiger-mode operation, 143–149, 144f, 149f, 186,
    189, 203–204
Germanium (Ge) detectors, 121, 126, 135–138,
    137f
Global ADCs, 53–54, 54f, 56f
Global shutter readout, 250, 252–253

H

HVCMOS detectors, 30–35, 32f, 34f, 35f
HVCMOS development, 33–34
HVCMOS processes, 257, 257f
HVCMOS sensors
    amplifiers, 38–39, 38f
    analog electronics for, 27–46
    bias resistance, 35–37, 37f
    continuous reset circuits, 39–40, 39f
    cross section of, 257f
    design details, 34–41
    discriminator, 40–41, 40f, 41f
    latch, 40–41, 41f
    new developments, 41
    passive shaper, 39–40, 40f
    resistive feedback, 39
    threshold control, 39–40, 40f
HVPixel chip, 33–34, 36f
HVPixelM detector, 33–35, 34f, 35f
Hybrid pixel detectors, 111, 269–271, 274–275;
    see also Pixel detectors
Hybrid semiconductor detectors, 266–269

I

Image sensors
    ADC architectures, 62
    ADC architectures for, 47–53, 66–67
    ADC comparisons, 62–66, 63t
    ADC efficiency, 62, 64f
    ADC requirements, 58–62
    ADC resolution, 58–60, 58f, 59f, 60f
    ADC topologies, 53–58
    area, 61

column-parallel ADCs, 54–55, 54f, 56f,
    109–110, 111f
cyclic ADCs, 52, 52f
design considerations, 62
$\Sigma\Delta$ ADCs, 53, 53f
figure of merit for, 64–67, 65t
global ADCs, 53–54, 54f, 56f
photon shot noise, 59–60, 59f, 60f
pipeline ADCs, 49, 50f
pixel-level conversion, 55–56, 56f
power consumption, 61–62
for radiation detection, 237–258
ramp ADCs, 49–51, 51f, 56f
random noise, 60–61
SAR ADCs, 51–52, 52f
slope ADCs, 49–51, 51f
speed, 61
Incremental sigma–delta (ISD) ADCs
    block scheme, 72f
    building blocks, 83–85
    cascode amplifier, 84, 84f
    coefficient scaling, 82–83, 88
    column-level readouts, 73–75, 74f
    column-parallel readouts, 71–72, 72f
    digital-to-analog converter, 75, 82–87, 87f
    first-order ISD ADC, 75–77, 75f, 76f
    higher-order ISD ADC, 77–88
    incremental sigma-delta ADCs, 71–88
    integrator, 83–85
    low-noise detectors and, 71–88
    PTC-based ADCs, 85–88
    quantizer, 85
    sample-and-hold block, 72–73, 72f
    second-order ISD ADC, 77–82, 78f, 79f,
        84–86, 87f
    switched capacitor summing circuit, 82–83, 83f
    third-order ISD ADC, 73, 80–81, 80f
    timing diagram, 76f, 87f
INMAPS, 29–30, 30f, 252, 252f
In-pixel electronics, 29–33, 252, 256–257; *see
    also* Pixel detectors
In-pixel intelligence, 252
Input counting rate (ICR), 123, 124f, 125f,
    129–130, 131f, 133f, 134f, 135f, 137f
Insulated gate field effect transistor (IGFET), 273
Integrated circuit (IC) readout system, 3, 20–24,
    21f
Integrated circuit (IC) shaper, 16, 16t
Ionization effects, 243–244
ISD ADCs; *see also* Incremental sigma–delta
    ADCs
    building blocks, 83–85
    cascode amplifier, 84, 84f
    coefficient scaling, 82–83, 88
    digital-to-analog converter, 75, 82–87, 85, 87f
    first-order ISD ADC, 75–77, 75f, 76f
    higher-order ISD ADC, 77–82, 84–88

integrator, 83–85
low-noise detectors, 71–88
PTC-based ADCs, 85–88
quantizer, 85
second-order ISD ADC, 77–82, 78f, 79f,
84–86, 87f
switched capacitor summing circuit, 82–83, 83f
third-order ISD ADC, 73, 80–81, 80f
Isolated n-well monolithic active pixel sensors
(INMAPS), 29–30, 30f, 252, 252f

K

KTC noise, 248–249, 248f, 251; *see also* Noise

L

Low voltage differential signalling (LVDS),
231–233, 256
Low-noise column readout circuit, 73–75
Low-noise detectors, 71–88; *see also* Noise

M

MAPS, 28–32, 28f, 29f, 30f, 32f, 46
Medipix detectors, 111–112, 273, 276
Metal–oxide semiconductor (MOS) transistor
amplification and, 18–19
analog signal processing and, 3–10
CMOS sensors and, 244–248, 246f, 248f
equivalent noise charge and, 4–6, 4f
level shifter dimensions, 18t
Metal–oxide-semiconductor field-effect transistor
(MOSFET)
amplification and, 18–19
analog signal processing and, 3–10
level shifter dimensions, 18t
OTA dimensions, 19t
radiation detection and, 246
silicon photomultipliers and, 224–225
transconductance amplifiers and, 231
Microchip, full-fabricated, 20, 20f
Microphotograph, 20, 20f
Modular pixel detectors, 265–279
Monolithic active pixel sensors (MAPS), 28–32,
28f, 29f, 30f, 32f, 46
Monolithic CMOS, 238–239
Mu3e, 41–42, 42f
Multichannel analyzer (MCA), 122–126, 132
MuPix, 42–46, 43f, 44f, 45f

N

N-channel metal oxide semiconductor (NMOS)
transistor
CMOS sensors and, 244, 244f, 251–252
CMOS-compatible technologies, 192

PMOS transistor and, 28–42, 29f, 30f, 32f,
37f, 104, 252
silicon photomultipliers and, 226
Noise
1/f noise, 3–6, 245, 248–249
analysis of, 2–6, 2f
column-level multiple-sampling ADCs,
73–75, 74f
of CSP–shaping filter system, 2–6, 2f
equivalent noise charge, 4–6, 10, 20–21, 225,
272–273
fixed-pattern noise, 48–54, 60–61, 61f, 66, 74,
88, 238, 273
flicker noise, 3–4, 249, 269
KTC noise, 248–249, 248f, 251
low-noise detectors, 71–88
noise performance, 187
noise sources, 246–249, 248f, 272–274
photon shot noise, 59–60, 59f, 60f, 85–86,
86f, 247
pixel detectors and, 272–274
random noise, 60–61, 249
reset noise, 248–249, 248f, 251
sensitivity, 190–192
signal/noise ratio, 2, 12, 39–40, 62, 85–88,
122, 153, 213, 231, 256–258
thermal noise, 3–7, 62, 73–75, 83–84, 123,
145, 247–249, 269, 273

O

Operational transconductance amplifiers (OTAs),
13–22, 15f–19f, 19t, 230; *see also*
Transconductance
Optical radiation detectors, 185–196
Optimization, 2–6
Output counting rate (OCR), 123, 124f, 125f,
129–130, 131f, 135f
Output signal undershoots, 22–23, 23f

P

Particle detectors, 1, 212, 265
P-channel metal oxide semiconductor (PMOS)
transistor
CMOS-compatible technologies, 192
NMOS transistor and, 28–42, 29f, 30f, 32f,
37f, 104, 252
transconductance amplifiers and, 19
Peak pileup, 123, 129, 130f, 135, 136f, 137f
Peaking time capability, 21–23, 23f
Peaking time regions (PTRs), 129–130
Phase blender, 104, 104f
Phase interpolator, 104
Phase-locked loop (PLL) circuit, 93, 99–100,
100f, 112–113, 113f
Photomultiplier tube (PMT), 142–143

Photomultipliers, 141–182, 203–233; *see also* Silicon photomultipliers
Photon capture, 147–148
Photon detection efficiency (PDE)
    measurement of, 173, 175f
    SiPM sensor parameters, 146, 150–152, 152f, 156, 167
    SiPM uniformity, 160–162
    TSV package, 170–173, 170f, 171f, 172f
Photon detection probability (PDP), 186–188, 192–196, 192f, 195f
Photon shot noise, 59–60, 59f, 60f, 85–86, 86f, 247; *see also* Noise
Photon transfer curve (PTC), 73, 85–88, 87f
Pipeline ADCs, 49, 50f
Pixel detectors
    3DIC technology, 271, 277–279, 277f, 278f
    application-specific integrated circuit readouts for, 267–279
    area interconnect, 270–272, 271f
    cooling, 274, 274f
    dead areas, 274–276, 275f
    flatness, 279
    hybrid detectors, 111, 269–271, 274–275
    in-pixel electronics, 29–33, 252, 256–257
    I/O connection, 275–276, 276f, 279
    modular detectors, 265–279
    Mu3e detectors, 41–42, 42f
    MuPix detectors, 42–46, 43f, 44f, 45f
    noise sources, 272–274
    smart pixels, 31–34, 36f
    stress, 279
    technology advances, 267–268, 271, 274, 277–279, 277f, 278f
    technology needs for, 265–279
    wire-bonded detectors, 265–266, 274–277, 275f, 276f, 277f
Pixel electronics, 30–34, 41–45, 42f, 45f, 257
Pixel readout cells, 42–45, 45f
Pixel TDC, 110–111
Positive-intrinsic-negative (PIN) photodiode, 142–143
Preamplifier–shaper structure, 2–6, 2f
Pulse shape and height analysis (PSHA), 126–129
Pulse shape correction (PSC), 129, 132
Pulse shape discrimination (PSD), 129, 132, 132f, 135, 136f, 137f
Pulse-processing techniques, 121–138; *see also* Digital pulse-processing techniques
Pulse-shrinking TDC, 101–102, 102f, 103f

R

Radiation detection
    ADC comparisons, 62–66
    ADC efficiency, 62, 64f
    ADC requirements, 58–62
    ADC resolution, 58–60, 58f
    ADC topologies, 53–58
    analog electronics for, 47–67
    applications, 256–258
    charge collection, 240–243
    CMOS image sensors for, 237–258, 239f
    design considerations, 62
    displacement effects, 244–245
    image sensor architectures, 47–53, 62, 66–67
    image sensors for, 237–258
    ionization effects, 243–244
    noise sources, 246–249
    pixel detectors and, 31
    radiation effects, 243–245
    radiation hardness, 243–246, 246f
    in silicon, 239–243
    state-of-the-art design, 62
    trends, 256–258
Radiation detectors
    CMOS image sensors, 237–258, 239f
    CMOS-compatible technologies, 185–196
    designing, 185–196
    optical radiation detectors, 185–196
    silicon radiation sensors, 237–258
Radiation effects, 243–245
Radiation hardness, 243–246, 246f
Radiation-detection IC, 20–24, 21f, 22f
Ramp ADCs, 49–51, 51f, 56f
Random noise, 60–61, 249; *see also* Noise
Reset circuits, 39–40, 39f
Reset noise, 248–249, 248f, 251; *see also* Noise
Resistive feedback, 39, 122, 131
Resistor-capacitor (R-C) delay, 108, 110f
Ring oscillator (RO), 99–100, 101f, 105–108, 105f, 106f, 112f
RLC network, 14–16, 14f, 16t
Rolling-shutter detectors, 33–34, 250, 252–253
Root mean square (RMS), 4–5, 96, 108, 213–214

S

Sample-and-hold (S&H) block, 72–73, 72f
SAR ADCs, 51–52, 52f
Scintillation crystal readouts, 141–182, 142f
Semiconductor detectors
    digital pulse-processing techniques for, 121–138, 122f
    gamma-ray semiconductor detectors, 121–138, 122f
    hybrid semiconductor detectors, 266–269
    X-ray semiconductor detectors, 121–138, 122f, 266–269, 268f
Set-reset (SR) flip-flops, 114, 115f
S-G shaping filter design, 10–13, 11f, 15f, 16f, 22–23
Shaping filter system, 6–13, 11f, 15f, 16f, 22–23
Shaping structure, 2–10, 2f

Sigma-delta ADCs, 71–88; *see also* Incremental
    sigma–delta ADCs
Signal/noise ratio (SNR); *see also* Noise
    description of, 2, 12
    HVCMOS sensors and, 39–40
    photon transfer curve and, 85–88
    pulse-processing techniques, 122
    radiation detection and, 62, 256–258
    silicon photomultipliers and, 213
    SiPM sensor parameters, 153
    transconductance amplifiers and, 231
Silicon photomultipliers (SiPMs)
    amplification, 147–150
    architecture examples, 225–233, 226f, 227f,
        228f, 230f, 232f, 233f
    charge-sensitive amplifier readout, 212–217,
        212f, 214f, 216f
    CRT, 179–180, 179f, 180f
    crystal performance, 178–179
    crystal readouts, 141–182, 142f
    current-mode readout, 222–225, 222f, 223f,
        224f
    energy resolution, 177–178, 177t, 179f,
        180–182, 181f
    front-end electronics for, 203–233
    gamma-ray detectors, 175–177
    Geiger-mode APD, 143–149, 144f, 149f
    history of, 143
    introduction of, 203–204
    light yield, 175–176, 176f
    microcells, 206–209, 206f
    models, 204–211, 207f, 209f, 210f, 214f, 215f,
        218f
    operation details, 143, 147–150
    output, 145–147, 147f, 207–210, 211f,
        215–217, 224, 233
    packaging, 165–172
    parameters, 150–160, 150t, 181, 210, 210t, 218
    photon absorption, 147–148, 148f
    photon capture, 147–148, 148f
    pulse output, 145–147, 147f
    readouts, 141–182, 142f, 207–208, 207f,
        212–225, 212f
    reliability, 172–173, 174t
    schematic of, 145f
    scintillation crystal readouts, 141–182, 142f
    sensor parameters, 150–160, 150t, 181, 218
    structure, 204–211, 205f, 206f
    timing measurements, 177–178
    uniformity, 160–164
    voltage-mode readout, 217–222, 218f, 219f,
        220f, 221f
Silicon radiation sensors, 237–258; *see also*
    Radiation detection
Single-photon avalanche diode (SPAD) detector,
    110–111, 112f, 186–190, 188f–191f

SiPM electronics, 203–233; *see also* Silicon
    photomultipliers
SiPM microcells, 206–209, 206f
SiPM model, 204–211, 207f, 209f, 210f, 214f,
    215f, 218f
SiPM operation, 147–150
SiPM packaging
    clear MLP, 167–172, 168f, 169f, 171f, 172f
    comparisons, 168t
    options, 166–167, 168t
    PDE, 160–162, 167, 170–172, 171f
    poured epoxy packages, 166–167, 166f, 172f
    TSV package, 170–173, 170f, 171f, 172f
SiPM reliability, 172–173, 174t
SiPM sensor parameters
    afterpulsing, 150, 158–159, 159f
    breakdown voltage, 150, 154–156
    crosstalk, 150, 156–157, 157f
    dark count rate, 150, 152–154, 153f, 154f
    dynamic range, 150, 159–160, 160f
    gain, 150, 154–156, 155f
    PDE, 146, 150–152, 152f, 156
    physical parameters, 150–160, 150t
SiPM structure, 204–211, 205f, 206f
SiPM uniformity
    breakdown voltage, 161, 162f, 164f
    dark current, 161–162, 163f
    optical current, 162–164, 163f
Slope ADCs, 49–51, 51f
Small-signal equivalent circuit, 8, 8f
Smart pixels, 33–34, 36f
Stochastic TDC, 116–117, 116f
Subgate delay resolution
    phase blender, 104, 104f
    phase interpolator, 104
    ring oscillators, 105–108, 105f, 106f
    Vernier TDC, 104–105, 104f

T

Tail pileup, 123, 129
TDC applications
    all-digital phase-locked loop, 112–113, 114f
    automobile applications, 110
    column-parallel ADCs, 109–110, 110f
    Medipix detectors, 111–112
    phase-locked loop, 112–113
    pixel TDC, 110–111
    single-photon avalanche diode detector,
        110–111, 112f
    Timepix detectors, 111, 112
TDC circuits; *see also* Time-to-digital converter
    advanced circuits, 114–117
    all-digital phase-locked loop, 112–113, 114f
    AMT chips, 108–109, 109f
    array oscillator, 105–106, 105f

counter, 98, 98f
custom chips, 107–109
delay-line TDC, 99–101, 99f, 100f, 102f
delay-locked loop circuit, 99, 100f, 105–106
field-programmable gate array, 105–107
phase-locked loop circuit, 93, 99–100, 100f,
    108, 112–113, 113f
pipelined TDC, 116, 116f
pulse-shrinking TDC, 101–102, 102f, 103f
resistor-capacitor delay, 108, 110f
ring oscillators, 99–100, 101f, 105–108, 105f,
    106f, 112f
set-reset flip-flops, 114, 115f
shift register, 98, 98f
stochastic TDC, 116–117, 116f
subgate delay resolution, 101–105
time difference amplifier, 114–117, 115f
time-to-amplitude converter, 98, 98f
timing diagram, 98f, 104f
TMC chips, 107–108, 108f
wave union method, 107, 107f
Telescopic cascode amplifier, 84f, 85
Thermal noise; see also Noise
    CMOS sensors and, 247–249
    column-level multiple-sampling ADCs, 73–75
    description of, 3–7
    photon transfer curve and, 83–84
    pixel detectors and, 269, 273
    radiation detection and, 62
    semiconductor detectors and, 123
    silicon photomultipliers and, 145
Time analog-to-digital converter (TAD), 110,
    112f
Time difference amplifier, 114–117, 115f

Timepix detectors, 111–112
Time-resolved imaging, 185–187
Time-to-amplitude converter (TAC), 98–100, 98f
Time-to-digital converter (TDC); see also TDC
    circuits
    conversion errors, 95–96, 95f, 96f
    measurement errors, 95–97, 95f, 96f, 97f, 98f
    operation mode, 92f, 93–95, 93f, 94f
    performance figures, 92f, 93–97, 93f, 94f
    timing diagram, 98f, 104f
TMC chips, 107–108, 108f
Transconductance, 3, 10, 38–39, 224, 245, 272
Transconductance amplifiers, 13–22, 15f–19f,
    19t, 231
Transconductance circuit, 19
Transconductance gain, 15
Triple-well MAPS, 29–32, 30f, 32f
TSV package, 170–173, 170f, 171f, 172f

V

Vernier TDC, 104–105, 104f
Voltage-mode readout, 217–222, 218f, 219f, 220f,
    221f

W

Wave union, 107, 107f
Wire-bonded detectors, 265–266, 274–277, 275f,
    276f, 277f

X

X-ray semiconductor detectors, 121–138, 122f

Printed and bound by CPI Group (UK) Ltd, Croydon, CR0 4YY

22/10/2024

01777613-0009